A Winning Bet In Nevada Baccarat

Edward Thorp & William Walden

Girard &
Stewart

Published by Girard & Stewart

ISBN: 978-1-62654-945-6

Printed in the U. S. A.

Foreword

By Jerome Montague

Nevada Baccarat, a widely played variation of Chemin de Fer (itself a direct variation of Baccarat), is played in numerous Nevada casinos and is popular throughout the gambling world.

Edward Thorp and William Walden give us a clear and definitive outline of how the game of Nevada Baccarat works by addressing its rules and more importantly how to understand probabilities. They illustrate when the side bet becomes the primary focus and provide a clear algorithmic guide to placing the most likely profitable bet. Where the lengthy technical explanation would be impractical in the fast-paced game of Nevada Baccarat, they give upper and lower bounds for the probabilities of financial ruin on the gambling floor. Finally, Thorp and Walden outline an equation to use for the player to know when it is time to stop betting on a given round.

While the rules of Baccarat can seem intimidating at first, once explained they become simple enough. In Baccarat, eight decks of cards are shuffled and placed into a wooden dealing box called a shoe. A card is drawn from the top of the deck and its value is noted. The number value of the card is then used to determine how many cards are to be subtracted from the deck and removed from play. Thus, if a nine is drawn, nine cards are taken out of the total deck. A Jack, Queen, or King is worth 10 cards.

Two players are then chosen to play as the Banker and the Player, respectively. The Banker represents the house, while the Player represents the players. Bets are placed on whether the Banker will win, the Player will win, or they will tie. The players who represent the Banker and the Player are not beholden to bet on themselves; they simply act out the game so that bets can be placed and gameplay runs smoothly.

Although the scoring in Nevada Baccarat is based on a simple point system, it makes for an exciting game. The ace is worth one point while the two through 10 cards are each worth their face value. The jack, queen, and king cards are each worth 10 points, yet given the way that Nevada Baccarat is scored (if a player surpasses nine points in a hand, 10 points are subtracted from the total), these cards are essentially worth a net of zero points. For example, if the Banker is dealt a five and a nine, his score is 14 minus 10 to equal four. At no point does the score ever become higher than nine. So whenever a face card is drawn, the 10 points it adds to your score is immediately subtracted. An easy way to think about this is that the Jack, Queen, and King cards are worth zero points.

The goal in Nevada Baccarat is to reach eight or nine before the other player. If either player reaches eight or nine points with their first two cards, the round is over and all bets are settled. This is known as a natural eight or natural nine, with a natural nine will always beating a natural eight. The only time that a natural eight or nine can occur is when the first two cards are drawn. If no natural eight or nine is drawn, cards are dealt until either the Player or the Banker reaches eight or nine. If all the cards of the deck are dealt with neither the Banker nor the Player reaching eight or nine, the round is over, and all players can make new bets. The deck is then reshuffled, and a new round begins.

Like most card games, Nevada Baccarat involves both luck and skill. Luck determines whether the game plays out in favor of the Banker or the Player. Skill becomes of utmost importance when it comes to betting. The highest paying bets are the side bets, which number four in total, and if successful, they pay out nine to one. These are placed on possible outcomes of the hand as opposed to simply whether the Banker or the Player will win.

Given the wealth of information presented in *A Winning Bet in Nevada Baccarat,* any serious or aspiring Nevada Baccarat players will benefit from understanding the use of applied mathematics to hone their betting strategy.

A Winning Bet in Nevada Baccarat

by

Edward Thorp
Department of Mathematics
New Mexico State University

and

William Walden
Los Alamos Scientific Laboratory*

CONTENTS

List of Tables

List of Figures

1. Introduction 1

2. Procedure in Baccarat 2

3. When the side bet is advantageous 5

4. The gambling system 7

5. The other side bets 21

6. Precise average casino advantage on each of the bets in Baccarat 35

7. The problem of gambler's ruin using the gambling system 37

8. The effects of approximating the optimal fixed fraction 44

9. Practical casino play 48

---

*This work was supported in part by the National Science Foundation, under Research Grant NSF-G 25058, and in part by the Atomic Energy Commission.

Part of the second author's doctoral dissertation, written at New Mexico State University, is based on the material in this paper.

10. When to stop playing in a given shoe 53

11. The distribution of f^* for various N 55

12. When the side bet on Banker natural nine is better than the main bets 58

13. The main bets 61

14. Applications to Blackjack 68

References 70

LIST OF TABLES

1. Nevada Baccarat rules 4

2.- 8. Betting fraction f^* of capital for various pairs (n, t) 11-17

9. The dependence of the exponential rate of growth G_{max}, and of the mean number of hours to double the initial capital, on the value of N. It is assumed that bets are made on Banker natural nine only 18

10. Comparison of f^{*B} and f^{*BP} 27

11. The probability of ruin when betting Banker natural nine only* 42

12. The probability of ruin when betting both Banker natural nine and Players natural nine 43

13. The effect on G of deviating from f^* 45

14. Precise amount to nearest dollar of joint bet that both the banker and the player have a natural nine (or eight) 50

15. Approximate bet sizes for V_0 = \$10,000, V_0 = \$20,000 and bets on the banker naturals only 52

16. Values of the ratio n/t for various choices of E, n and N 54-55

17. The distribution of f^* for various N, when the Banker natural nine is bet 57

18. When a (minimum, unfavorable) side bet on Banker natural nine gives the same rate of growth as an equal bet on the Banker or The Player 59

19. When Banker natural nine is to be preferred over the main bets, using the criterion $p_B = p_P > 0.1$ 61

20. Dependence of main bet probabilities on deck composition 63-64

21. Main bet probabilities for randomly chosen 13 card subsets.

LIST OF FIGURES

1. The dependence of the exponential rate of growth G_{max}, and of the mean number of hours to double the initial capital, on the value of N. It is assumed that bets are made on Banker natural nine only 19

2. Illustrating the effect on G of deviating from f^* 46

3. The relation of f_c to f^* when $0 \leq f^* \leq 0.14$ 47

1. Introduction.

The games of Baccarat and Chemin de Fer are well known gambling games played for high stakes in several parts of the world. Chemin de Fer is a direct modification of Baccarat. Descriptions and background for the two games may be found in [3, 7]. Earlier studies of Baccarat appear in [1, 5, 6]. A few Nevada casinos offer a modification of Chemin de Fer which we refer to as Nevada Baccarat. The rules, structure and format of the three games have strong similarities. We studied Nevada Baccarat because the casinos where it is played are readily accessible. Our techniques can be carried over to the other two games. From now on the term Baccarat refers exclusively to Nevada Baccarat, except when noted.

Our study began with the observation that Nevada Baccarat is vaguely similar to the game of Blackjack, or Twenty-One. The fact that practical winning strategies for Twenty-One have recently been discovered [8, 9, 10] suggested there might also be practical winning strategies for Baccarat.

In contrast to the situation in Twenty-One, we found that there are no practical winning strategies for the main part of the game, i.e. for the even money Banker and Player bets. The evidence for this is given in the next to last section of this paper.

During this work we learned that the game sometimes has certain associated sidebets. In the principal part of this paper we show that there is a practical winning strategy for these side bets.

A player following this strategy increases his capital exponentially. In practice, the mean doubling time is twenty five hours, or so. The probability of eventual ruin is very small. It depends on the player's initial capital and in practice can be made quite small. The strategy depends on the fact that during the course of the game the composition of the playing deck changes. The player's probability of winning a side bet is advantageous for some of these deck compositions. These probabilities and their frequency of occurrence are determined. Kelly's criterion [4] determines the amount to be bet in a favorable situation.

The general idea strongly resembles the card-counting and money management principles developed in the winning strategy for Blackjack described in [8, 9, 10]. In the last section of this paper we show how our analysis of the side bet in Baccarat applies to Blackjack. It turns out that sections 4, 7, 8 and 11 can be imitated for Blackjack, often step by step. The result is a fairly detailed analysis of the Kelly gambling system applied to Blackjack.

2. Procedure in Baccarat.

We give those details of Baccarat which we need. Variations from the format we describe are insignificant for our purposes unless otherwise noted. To begin the game, eight decks of cards are shuffled and a joker is placed face up near the end. The cards are then put into a wooden dealing box called a shoe. The first card is exposed, and its value is noted, face cards being counted as tens. Then this number of cards are discarded, or "burned".

The table has twelve seats, occupied by an assortment of customers and shills. We refer to them indiscriminately as "players". There are two principal bets, called "Banker" and "Players". Any player may make either of these bets before the beginning of any round of play, or "coup".

To begin the evening's play, two of the players are singled out. One is termed The Banker and the other is termed The Player. The Banker retains the shoe and deals as long as the bet "Banker" (which we also refer to as a bet on The Banker) does not lose. When the bet "Players" (which we also refer to as a bet on The Player) wins the shoe moves to the player on the right. This player now becomes The Banker. If the coup is a tie, the players are allowed to alter their bets in any manner they wish. The same Banker then deals another coup. The Player is generally chosen to be that player, other than the one who is temporarily The Banker, who has the largest bet on Players. We have not noticed an occasion where there were no bets on Players, because when we have played there have always been shills in the game and they generally bet on The Player (except when acting as The Banker, when they generally bet on The Banker).

To begin a coup, The Banker and The Player are dealt two cards each. The cards Ace through Ten are each worth their face value and the cards Jack through King are each worth ten points. A hand is evaluated as the sum modulo ten of his cards.

After The Banker and The Player each receive two cards, the croupier faces their hands. If either two - card total equals 8 or 9 (termed a natural 8 or natural 9, as the case may be), all bets are settled at once.

If neither The Player nor The Banker have a natural, The Player and The Banker then draw or stand according to the set of rules in Table 1.

Table 1. Nevada Baccarat Rules

Player

having

0-5	draws a card
6-7	stands
8-9	turns cards over

Banker

having	draws when The Player draws	does not draw when The Player draws
0	none, 0-9	
1	none, 0-9	
2	none, 0-9	
3	none, 0-7,9	8
4	none, 2-7	0,1,8,9,
5	none, 4-7	0-3,8,9,
6	6,7	none, 0-5,8,9,
7	stands	stands
8	turns cards over	turns cards over
9	turns cards over	turns cards over

The high hand wins. If the hands are equal, there is a tie and no money changes hands. Players are then free to change their bets in any desired manner. If the coup being played is complete when the joker is reached, the shoe ends and the cards are reshuffled. Otherwise the coup is played out to completion and then the shoe ends and the cards are reshuffled. However, the casino may reshuffle the cards at any time.

Winning bets on The Player are paid at even money. Winning bets on The Banker are paid 0.95 of the amount bet. The 5% tax which is imposed on what otherwise would have been an even-money pay-off is called "vigorish". It is the source of the casino's profit on the main bets in the game.

In addition to the bets on The Banker and The Player there are four side bets, some or all of which may occur. They are that The Banker receives a natural 9, The Banker receives a natural 8, The Player receives a natural 9, and The Player receives a natural 8. They each pay 9 to 1. The development that follows will be concerned with the first of these bets. The analysis of all four, whether taken separately or in combination, can then be deduced.

We tested our strategy successfully in the casinos. We studied the side bet in two particular casinos, which we refer to as casinos A and B. In casino A, the limits on the main bet were \$5 - \$2000. Bets on the natural ranged from \$5 to \$200 on natural 8 and an additional \$5 to \$200 on natural 9. The bet on either natural could be divided between The Player and The Banker in desired proportions. In casino B, the limits on the main bet were \$20 - \$2000 and bets on Banker naturals only were allowed. The limits were \$20 - \$2000 on each.

3. When the side bet is advantageous.

Let n be the number of cards in the shoe that thus far have not been used in play. Let t denote the number of these n cards with the value nine. Suppose the n-t other cards are typically distributed. This means that we assume that they have been selected at random from the 384 non-nines which were originally in the complete eight decks. Let two cards d_1 and d_2 be drawn from the n cards. By considering cases, we find the probability that The Banker's first two dealt cards total nine. Clearly the side bet has positive expectation if and only if this probability exceeds 0.1.

(a) $d_1 = 9$, $d_2 = 0$

$\text{Prob}(d_1 = 9) = t/n$. $\text{Prob}(d_2 = 0) = (n-t)/[3(n-1)]$.

If $d_1 = 9$ and $d_2 = 0$ then $\text{Prob}(d_1+d_2 = 9) = 1$

$\text{Prob}(d_1 = 9 \text{ and } d_2 = 0 \text{ and } d_1+d_2 = 9) = [t(n-t)]/[3n(n-1)$

(b) Similarly,

$\mathrm{Prob}(d_1 = 0 \text{ and } d_2 = 9 \text{ and } d_1+d_2 = 9) = [t(n-t)]/[3n(n-1)]$

(c) $d_1 \neq 9,\ d_2 \neq 9$

$\mathrm{Prob}(d_1 \neq 9) = (n-t)/n.$ $\mathrm{Prob}(d_2 \neq 9) = (n-t-1)/(n-1)$

If $d_1 \neq 9$ and $d_2 \neq 9$ then $\mathrm{Prob}(d_1+d_2 = 9) = [8(32)(32)]/[(384)(383)]$.

$\mathrm{Prob}(d_1 \neq 9 \text{ and } d_2 \neq 9 \text{ and } d_1 + d_2 = 9)$

$$= [8(32)(32)(n-t)(n-t-1)]/[(384)(383)n(n-1)].$$

By combining the three cases above, the probability of obtaining a total of nine by drawing two cards is given by:

(1) $$p_{n,t} = [2(n-t)(32n+351t-32)]/[1149n(n-1)].$$

We next determine how the favorable side bets are distributed. Given n cards, the probability that t of these cards are nines is given by

$$a_{n,t} = [32!384!n!(416-n)!]/[416!t!(n-t)!(32-t)!(384-n+t)!].$$

We also have the following recursion formulas:

$$a_{n,t} = a_{n-1,t}\ [n(385-n+t)]/[(n-t)(417-n)].$$

$$a_{n,t} = a_{n,t-1}\ [(n-t+1)(33-t)]/[t(384-n-t)]$$

Let N' be the total number of cards which are not used during play, i.e. N' is the sum of the unseen burned cards and the cards left in the shoe when the house decides to shuffle. Then $416-N'$ cards are actually dealt. Thus the number n' of cards which as yet have not been used at some instant during play of a shoe ranges from 416 to $416-N' + 1$.

Either 4, 5 or 6 cards are used during a coup. Thus the number n of cards which as yet have not been used just before the play of some coup begins in a shoe, is one of the numbers 416, 412 to 410, 408 to $(N' + 4)$ and all these values can be attained. It is n which is significant for our purposes. We define $N = (N' + 4)$, the least value which n can attain during the play of the shoe. Of course N generally varies from one run through the eight decks (also termed a shoe) to another.

We assume that the various values of n between 416 and (416-N) occur with equal probability. This is only an approximation to what actually occurs but it appears to introduce negligible errors in our calculations. With this assumption, the probability that the first card which is drawn in a given coup during the play of a shoe is drawn from n cards, is 1/(416-N). Therefore the probability that the draw of the first card to a hand is from n cards, t of which are nines, is given to good approximation by: $c_{n,t} = a_{n,t}/(416-N)$. Complete tables of $c_{n,t}$ were obtained from this relation, the recursion formulas for $a_{n,t}$, and a high speed computer. Likewise the values of $p_{n,t}$ for all possible n and t were determined via the computer. It turns out that when N is fairly small, say 20 or 30, favorable side bets can be placed 20 per cent of the time. The advantage on (mathematical expectation of) these side bets runs as high as 89 per cent when N is 20.

4. The gambling system.

Now the gambling system used will be described. To get an idea of what is involved, consider a coin toss game in which p, the probability of our winning, is greater than 1/2. Suppose that we have V_0 units of capital initially and V_t units before trial t. How should we bet? Since $p > 1/2$, the expectation is positive so we might bet to maximize our expectation. But this means that we should bet our entire current capital of V_t units at each trial. Consequently, with probability 1 we can expect to be ruined.

Alternately we might bet just 1 unit, i.e. the minimum bet allowed, at each trial. This generally makes the ruin probability small compared with many other conceivable betting schemes. However, one's capital grows rather slowly.

An interesting compromise is explored by Kelly in [1]. He assumes that money is infinitely divisible and that the player bets a fixed fraction f of his capital at each trial. It turns out that there is a fraction f* with the interesting property that if two players compete, one using f and the other f*, the probability that $V_t(f^*) > V_t(f)$ tends to 1 as t tends to ∞. There also is a critical fraction $f_c > f^*$ such that if $f > f_c$, then given $\varepsilon > 0$ and $\delta > 0$, there is a $t(\varepsilon,\delta)$ such that $t > t(\varepsilon,\delta)$ implies that the probability that $V_t(f) < \varepsilon$ is greater than $1-\delta$. For the coin toss game, $f^* = p - q$ and f_c is the real root between zero and 1 of the equation $(1+f)^p (1-f)^q = 1$.

In practice, the system of Kelly must be modified to fit reality, e.g. bets are integral multiples of a minimum unit. Because integral bets only are allowed, one can at best generally only approximate the bets f^*V_t called for by Kelly. Further, when V_t is small, f^*V_t may be much less than one unit and reasonable approximation is no longer possible. The player must either bet nothing or much more than f^*V_t.

Now we develop the Kelly ideas for our specific problem. Let $p_k > 1/10$, $1 \leq k \leq K$, be a probability that is known to occur with probability c_k. When p_k occurs, suppose the player bets a fraction f_k of his capital. If t bets are made there are t_k bets of type k with w_k wins and l_k losses, for $1 \leq k \leq K$. Denote the player's capital after t bets by V_t, and his initial capital by V_0.

Then $V_t = V_0 \prod_{k=1}^{K} (1+9f_k)^{w_k} (1-f_k)^{l_k}$.

Define the exponential rate of growth, G, by:

(2) $G = \lim_{t\to\infty}[(\log_2(V_t/V_0))/t]$.

Then $G = \lim_{t\to\infty}[(\log_2(\prod_{k=1}^{K}(1+9f_k)^{w_k}(1-f_k)^{l_k}))/t]$

$$= \lim_{t\to\infty}[(\sum_{k=1}^{K} w_k\log_2(1+9f_k)+l_k\log_2(1-f_k))/t]$$

$$= \lim_{t\to\infty}\sum_{k=1}^{K}[(w_k/t_k)(t_k/t)\log_2(1+9f_k)+(l_k/t_k)(t_k/t)\log_2(1-f_k)]$$

(3) $$= \sum_{k=1}^{K} c_k[p_k\log_2(1+9f_k)+(1-p_k)\log_2(1-f_k)]$$

By the strong law of large numbers, V_t tends to be very close to $V_0 2^{t/(1/G)}$ as $t \to \infty$. Note that $1/G$ will be the mean number of bets required to double the initial capital V_0. We wish to maximize G.

$$\frac{\partial G}{\partial f_k} = c_k\log_2 e[(9p_k)/(1+9f_k) - (1-p_k)/(1-f_k)].$$

Hence $\frac{\partial G}{\partial f_k} = 0$ iff $9p_k/(1+9f_k) = (1-p_k)/(1-f_k)$.

Solving for f_k gives:

(4) $f_k = (10p_k-1)/9.$

Also $1+9f_k = 10p_k$ and $1-f_k = 10(1-p_k)/9$ so that:

(5) $$G_{max} = \sum_{k=1}^{K} c_k[p_k\log_2(10p_k) + (1-p_k)\log_2(10(1-p_k)/9)].$$

All the formulas needed to construct the strategy are now available. Certain values of $p_{n,t}$ satisfy $p_{n,t} > 1/10$. These probabilities occur with probability $c_{n,t}$. For these pairs (n,t), $f_{n,t} = (10p_{n,t}-1)/9$. If $p_{n,t} \leq 1/10$, let $f_{n,t} = 0$, corresponding to no bet. Thus a table can be computed telling the player for each possible pair (n,t) what fraction of his initial capital should be wagered. However, such a table is too large to be practical for actual betting. Fortunately, examination of $f_{n,t}$ shows that for certain values of n, f_{n_1,t_1} is approximately equal to f_{n_2,t_2} if

$n_1/t_1 = n_2/t_2$. These results are given in tables 2-6. For $5 \leq n \leq 19$ tables 7 and 8 actually list n, t and $f_{n,t}$. Hence these seven tables give the strategy for betting that the banker has a total of nine on his first two cards. The use of this strategy in actual play will be described later.

The blank spaces in Tables 2-8 correspond to impossible situations where $t > 32$ would be required.

Table 9 lists N and $1/G_{max}$, showing the mean number of bets and number of hours necessary to double the initial capital for various reshuffling points, assuming 100 hands are played per hour. This is almost exactly the mean observed rate in the casinos.

TABLE 2. Betting fraction f* of capital for 1.3 ≦ n/t ≦ 3.4

	n							
n/t	20	25	30	40	50	65	80	≧100
1.3	.029	.028	.027	.026				
1.4	.052	.050	.049	.048				
1.5	.068	.066	.065	.064				
1.6	.079	.077	.076	.075	.074			
1.7	.087	.085	.084	.083	.082			
1.8	.092	.091	.089	.088	.087			
1.9	.096	.094	.093	.091	.090			
2.0	.098	.096	.095	.093	.093			
2.1	.099	.097	.096	.094	.094	.093		
2.2	.099	.098	.096	.095	.094	.093		
2.3	.099	.097	.096	.094	.094	.093		
2.4	.098	.096	.095	.094	.093	.092		
2.5	.097	.095	.094	.093	.092	.091	.091	
2.6	.096	.094	.093	.091	.090	.090	.089	
2.7	.094	.092	.091	.090	.089	.088	.088	
2.8	.092	.091	.089	.088	.087	.086	.086	
2.9	.090	.089	.088	.086	.085	.085	.084	
3.0	.089	.087	.086	.084	.084	.083	.082	
3.1	.087	.085	.084	.082	.082	.081	.081	
3.2	.085	.083	.082	.081	.080	.079	.079	.078
3.3	.082	.081	.080	.079	.078	.077	.077	.076
3.4	.080	.079	.078	.077	.076	.075	.075	.074

TABLE 3. Betting fraction f* of capital for $3.5 \leq n/t \leq 5.6$

				n					
n/t	20	25	30	40	50	65	80	100	≧130
3.5	.079	.077	.076	.075	.074	.073	.073	.073	
3.6	.077	.075	.074	.073	.072	.071	.071	.071	
3.7	.075	.073	.072	.071	.070	.070	.069	.069	
3.8	.073	.071	.070	.069	.068	.068	.067	.067	
3.9	.071	.069	.068	.067	.066	.066	.065	.065	
4.0	.069	.067	.067	.065	.065	.064	.064	.063	
4.1	.067	.066	.065	.064	.063	.062	.062	.062	.061
4.2	.065	.064	.063	.062	.061	.061	.060	.060	.060
4.3	.063	.062	.061	.060	.060	.059	.059	.058	.058
4.4	.062	.060	.060	.058	.058	.057	.057	.057	.056
4.5	.060	.059	.058	.057	.056	.056	.055	.055	.055
4.6	.058	.057	.056	.055	.055	.054	.054	.053	.053
4.7	.057	.056	.055	.054	.053	.053	.052	.052	.052
4.8	.055	.054	.053	.052	.052	.051	.051	.050	.050
4.9	.054	.053	.052	.051	.050	.050	.049	.049	.049
5.0	.052	.051	.050	.049	.049	.048	.048	.048	.047
5.1	.051	.050	.049	.048	.047	.047	.047	.046	.046
5.2	.049	.048	.047	.047	.046	.045	.045	.045	.045
5.3	.048	.047	.046	.045	.045	.044	.044	.044	.043
5.4	.047	.045	.045	.044	.043	.043	.043	.042	.042
5.5	.045	.044	.043	.043	.042	.042	.041	.041	.041
5.6	.044	.043	.042	.041	.041	.040	.040	.040	.040

TABLE 4. Betting fraction f* of capital for $5.7 \leqq n/t \leqq 7.8$

					n					
n/t	20	25	30	40	50	65	80	100	130	≥180
5.7	.043	.042	.041	.040	.040	.039	.039	.039	.038	.038
5.8	.041	.040	.040	.039	.038	.038	.038	.037	.037	.037
5.9	.040	.039	.039	.038	.037	.037	.037	.036	.036	.036
6.0	.039	.038	.037	.037	.036	.036	.035	.035	.035	.035
6.1	.038	.037	.036	.036	.035	.035	.034	.034	.034	.034
6.2	.037	.036	.035	.034	.034	.034	.033	.033	.033	.033
6.3	.036	.035	.034	.033	.033	.033	.032	.032	.032	.032
6.4	.035	.034	.033	.032	.032	.031	.031	.031	.031	.031
6.5	.034	.033	.032	.031	.031	.031	.030	.030	.030	.030
6.6	.033	.032	.031	.030	.030	.030	.029	.029	.029	.029
6.7	.032	.031	.030	.029	.029	.029	.028	.028	.028	.028
6.8	.031	.030	.029	.028	.028	.028	.027	.027	.027	.027
6.9	.030	.029	.028	.028	.027	.027	.026	.026	.026	.026
7.0	.029	.028	.027	.027	.026	.026	.026	.025	.025	.025
7.1	.028	.027	.026	.026	.025	.025	.025	.025	.024	.024
7.2	.027	.026	.026	.025	.025	.024	.024	.024	.024	.023
7.3	.026	.025	.025	.024	.024	.023	.023	.023	.023	.023
7.4	.025	.024	.024	.023	.023	.022	.022	.022	.022	.022
7.5	.024	.023	.023	.022	.022	.022	.021	.021	.021	.021
7.6	.023	.023	.022	.022	.021	.021	.021	.020	.020	.020
7.7	.022	.022	.022	.021	.020	.020	.020	.020	.020	.019
7.8	.022	.021	.021	.020	.020	.019	.019	.019	.019	.019

TABLE 5. Betting fraction f* of capital for $7.9 \leq n/t \leq 9.9$

	n										
n/t	20	25	30	40	50	65	80	100	130	180	≥240
7.9	.021	.020	.020	.019	.019	.019	.018	.018	.018	.018	.018
8.0	.020	.020	.019	.019	.018	.018	.018	.018	.017	.017	.017
8.1	.019	.019	.018	.018	.018	.017	.017	.017	.017	.016	.016
8.2	.018	.018	.018	.017	.017	.017	.016	.016	.016	.016	.016
8.3	.018	.018	.017	.017	.016	.016	.016	.015	.015	.015	.015
8.4	.017	.017	.016	.016	.016	.015	.015	.015	.015	.014	.014
8.5	.016	.016	.016	.015	.015	.015	.014	.014	.014	.014	.014
8.6	.016	.016	.015	.015	.014	.014	.014	.014	.013	.013	.013
8.7	.015	.015	.014	.014	.014	.013	.013	.013	.013	.013	.013
8.8	.014	.014	.014	.013	.013	.013	.012	.012	.012	.012	.012
8.9	.014	.013	.013	.013	.012	.012	.012	.012	.012	.011	.011
9.0	.013	.013	.012	.012	.012	.011	.011	.011	.011	.011	.011
9.1	.013	.012	.012	.011	.011	.011	.011	.011	.010	.010	.010
9.2	.012	.012	.011	.011	.011	.010	.010	.010	.010	.010	.010
9.3	.012	.011	.011	.010	.010	.010	.010	.009	.009	.009	.009
9.4	.011	.010	.010	.010	.009	.009	.009	.009	.009	.009	.008
9.5	.011	.010	.010	.009	.009	.009	.008	.008	.008	.008	.008
9.6	.010	.009	.009	.009	.008	.008	.008	.008	.008	.008	.007
9.7	.010	.009	.009	.008	.008	.008	.007	.007	.007	.007	.007
9.8	.009	.008	.008	.008	.007	.007	.007	.007	.007	.007	.006
9.9	.009	.008	.008	.007	.007	.006	.006	.006	.006	.006	.006

TABLE 6. Betting fraction f* of capital for $10.0 \leq n/t \leq 11.6$

	n										
n/t	20	25	30	40	50	65	80	100	130	180	≧240
10.0	.008	.007	.007	.007	.006	.006	.006	.006	.006	.006	.005
10.1	.007	.007	.007	.006	.006	.006	.005	.005	.005	.005	.005
10.2	.007	.006	.006	.006	.005	.005	.005	.005	.005	.005	.004
10.3	.007	.006	.006	.005	.005	.005	.005	.004	.004	.004	.004
10.4	.006	.005	.005	.005	.004	.004	.004	.004	.004	.004	.004
10.5	.006	.005	.005	.004	.004	.004	.004	.003	.003	.003	.003
10.6	.005	.004	.004	.004	.004	.003	.003	.003	.003	.003	.003
10.7	.004	.004	.004	.003	.003	.003	.003	.003	.003	.002	.002
10.8	.004	.003	.003	.003	.003	.002	.002	.002	.002	.002	.002
10.9	.004	.003	.003	.003	.002	.002	.002	.002	.002	.001	.001
11.0	.003	.003	.003	.002	.002	.002	.001	.001	.001	.001	.001
11.1	.002	.002	.002	.002	.001	.001	.001	.001	.001	.001	.001
11.2	.002	.002	.002	.001	.001	.001	.001	.001	.000	.000	.000
11.3	.002	.002	.001	.001	.001	.001	.000	.000	.000	.000	.000
11.4	.002	.002	.001	.001	.000	.000	.000	.000	.000	.000	.000
11.5	.001	.001	.001	.000	.000	.000	.000	.000	.000	.000	.000
11.6	.000	.000	.000	.000	.000	.000	.000	.000	.000	.000	.000

TABLE 7.[†] Betting fraction f* of capital for $5 \leq n \leq 12$

	n							
t	5	6	7	8	9	10	11	12
1	.074	.053	.038	.037	.019	.012	.006	.002
2	.129	.111	.094	.080	.069	.059	.050	.043
3	.117	.123	.118	.109	.099	.090	.082	.074
4	.037	.090	.109	.113	.111	.107	.101	.094
5		.012	.068	.093	.104	.108	.108	.104
6				.049	.079	.094	.102	.104
7					.034	.065	.084	.094
8						.021	.053	.074
9							.011	.043
10								.002

[†] A square in the table corresponds to a possible situation if and only if $t \leq n$. If a square which corresponds to a possible situation is blank, the appropriate entry is 0.000.

TABLE 8.[†] Betting fraction f* of capital for $13 \leq n \leq 19$

t	n = 13	14	15	16	17	18	19
1	.000	.000	.000	.000	.000	.000	.000
2	.036	.031	.026	.022	.018	.014	.011
3	.067	.060	.054	.049	.044	.040	.036
4	.088	.082	.076	.071	.065	.061	.056
5	.101	.096	.091	.087	.082	.077	.073
6	.104	.103	.100	.097	.093	.089	.086
7	.100	.102	.102	.101	.099	.097	.094
8	.086	.094	.098	.100	.101	.100	.099
9	.064	.078	.088	.094	.097	.099	.100
10	.033	.055	.071	.081	.089	.093	.096
11		.025	.047	.063	.075	.083	.089
12			.017	.040	.056	.069	.078
13				.010	.033	.050	.063
14					.004	.026	.044
15							.021

[†]A square in the table corresponds to a possible situation if and only if $t \leq n$. If a square which corresponds to a possible situation is blank, the appropriate entry is 0.000.

TABLE 9. The dependence of the exponential rate of growth G_{max}, and of the mean number of hours to double the initial capital, on the value of N. It is assumed that bets are made on Banker natural nine only.

N	$G_{max} \cdot 10^{-4}$	HOURS TO DOUBLE CAPITAL
5	3.052	33
6	2.821	35
7	2.637	38
8	2.485	40
9	2.358	42
10	2.247	45
15	1.826	55
20	1.542	65
25	1.336	75
30	1.173	85
35	1.043	96
40	0.934	107
45	0.843	119
50	0.764	131
60	0.635	158
70	0.533	188
80	0.451	222
90	0.384	260
100	0.328	305
150	0.149	671
200	0.063	1597
300	0.004	26712
363	0	

Figure 1. The dependence of the exponential rate of growth G_{max}, and of the mean number of hours to double the initial capital, on the value of N. It is assumed that bets are made on Banker natural nine only.

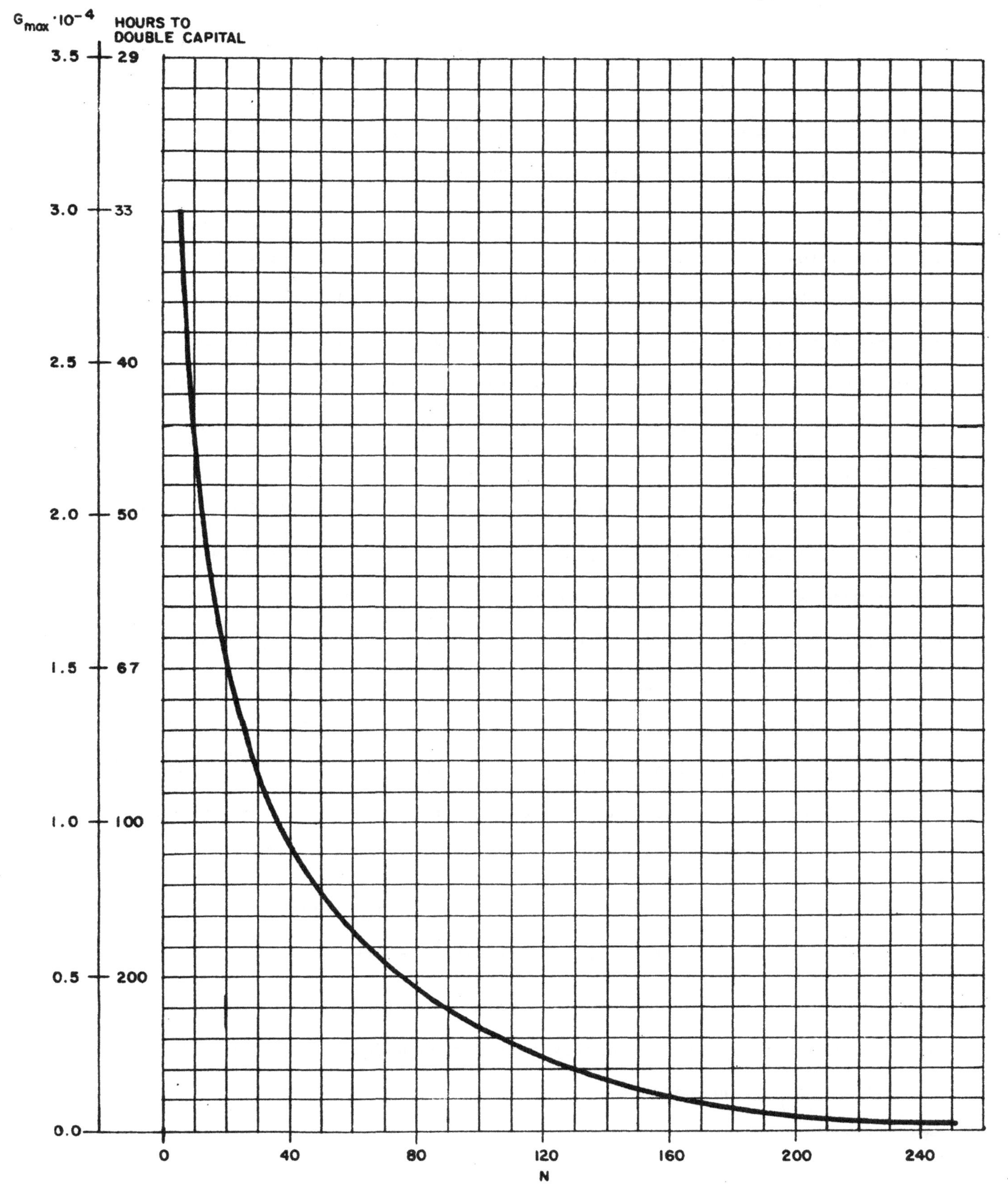

In casino A, N varied from 20 to 50, with a mean of 30 - 35. Table 9 shows us that our capital should double in about 90 hours, betting Banker natural nine only. Since "waiting bets" must also be made and since only approximate optimal fractions can be bet in favorable situations, an estimate of 100 hours for the mean doubling time is more realistic. Casino A allowed all four bets on the naturals. As we'll see below, the optimal betting fraction is approximately the sum of the separate fractions for each natural bet and the growth rate is approximately the sum of the growth rates. Further, these numbers are all approximately equal. Thus, the mean doubling time when all four naturals are bet is about 25 hours at casino A.

At casino B, N varied from 6 to 16, with a mean of about 11. Our corresponding estimate of mean doubling time is 50 hours when only the Banker natural nine is bet. Casino B allowed Banker natural eight bets also but forbid bets on the Player naturals. Thus mean doubling time was (coincidentally) again about 25 hours.

In the forty hours of play which we were permitted in casinos A and B, we happened to experience almost exactly the theoretical growth rate.

Some insight into the structure of tables 2 - 8 can be obtained by rewriting the expression for $p_{n,t}$. We have, after a brief calculation,

$$p_{n,t} = [2(n-t)(32n+351t-32)]/[1149n(n-1)] \tag{6}$$
$$= [2(r-1)/1149r][32+(351/r)(1+1/(n-1))]$$

where $r = n/t$. Therefore for a fixed value of r (we call r "the ratio"), $p_{n,t}$ and hence $f_{n,t}$, will gradually increase as n decreases towards N. We call this gradual increase in $f_{n,t}$ as n decreases, for fixed ratio r, as "the end effect".

5. The other side bets.

Now consider the bet that the banker has a total of eight on his first two cards. A treatment similar to that just given could be applied, yielding a set of tables for this bet. However it can be shown that such a set of tables agree, to a high degree of approximation, with those given for the first bet. The only formula changed is that for $p_{n,t}$. To distinguish, the notation $p^8_{n,t}$ and $p^9_{n,t}$ will be used. Suppose $|p^9_{n,t} - p^8_{n,t}| < \varepsilon$ and $p^9_{n,t} > 1/10$, $p^8_{n,t} > 1/10$. Then

$$|f^9_{n,t} - f^8_{n,t}| = |[(10p^9_{n,t}-1)/9]-[(10p^8_{n,t}-1)/9]|$$

$$= |(10/9)(p^9_{n,t} - p^8_{n,t})|$$

$$= 10/9|p^9_{n,t} - p^8_{n,t}| < 10\varepsilon/9$$

Now suppose $|p^9_{n,t} - p^8_{n,t}| < \varepsilon$ and $p^9_{n,t} > 1/10$, $p^8_{n,t} \leq 1/10$.

Then $f^8_{n,t} = 0$ and $p_{n,t} < 1/10 + \varepsilon$

Hence $f^9_{n,t} = (10p^9_{n,t}-1)/9 < (1+10\varepsilon-1)/9 = 10\varepsilon/9$

Similarly if $p^9_{n,t} \leq 1/10$, $p^8_{n,t} > 1/10$, then $f^9_{n,t} = 0$

and $f^8_{n,t} < 10\varepsilon/9$. Therefore if $|p^9_{n,t} - p^8_{n,t}| < \varepsilon$ then

$|f^9_{n,t} - f^8_{n,t}| < 10\varepsilon/9$. By computing ε, we shall show that

$|f^9_{n,t} - f^8_{n,t}| < 0.0005$ so the tables are the same for the second bet to within 0.0005. Therefore, considering the effects of rounding off, they agree to within 0.001.

Let n be as before. Of these n cards, denote by t the number of eights. As before suppose the n-t other cards are typically distributed.

Let two cards d_1 and d_2 be drawn from the n cards, and consider these cases:

(a) $d_1 = 8, d_2 = 0$

$\text{Prob}(d_1 = 8) = t/n$. $\text{Prob}(d_2 = 0) = (n-t)/[3(n-1)]$.

If $d_1 = 8$ and $d_2 = 0$ then $\text{Prob}(d_1+d_2 = 8) = 1$.

$\text{Prob}(d_1 = 8 \text{ and } d_2 = 0 \text{ and } d_1+d_2 = 8) = [t(n-t)]/[3n(n-1)]$.

(b) $d_1 = 0, d_2 = 8$

$\text{Prob}(d_1 = 0) = (n-t)/(3n)$. $\text{Prob}(d_2 = 8) = t/(n-1)$.

If $d_1 = 0$ and $d_2 = 8$ then $\text{Prob}(d_1+d_2 = 8) = 1$.

$\text{Prob}(d_1 = 0 \text{ and } d_2 = 8 \text{ and } d_1+d_2 = 8) = [t(n-t)]/[3n(n-1)]$.

(c) $d_1 \neq 8, d_2 \neq 8$

$\text{Prob}(d_1 \neq 8) = (n-t)/n$. $\text{Prob}(d_2 \neq 8) = (n-t-1)/(n-1)$.

If $d_1 \neq 8$ and $d_2 \neq 8$ then:

$\text{Prob}(d_1+d_2 = 8) = [6(32)(32)+2(32)(31)]/[(384)(383)]$.

$\text{Prob}(d_1 \neq 8 \text{ and } d_2 \neq 8 \text{ and } d_1+d_2 = 8)$

$$= [6(32)(32)+2(32)(31)](n-t)(n-t-1)/[(384)(383)n(n-1)]$$

By combining the three cases above, the probability of a total of eight by drawing two cards is given by:

$$p^8_{n,t} = [(n-t)(127n+1405t-127)]/[2298n(n-1)].$$

Then $p^9_{n,t} - p^8_{n,t} = [(n-t)(n-t-1)]/[2298n(n-1)]$

$$= (1/2298)[(n-t)/n][(n-t-1)/n]$$

$$< 1/2298.$$

Therefore $|f^9_{n,t} - f^8_{n,t}| < (10/9)/2298 < 0.0005$

Now consider the bet that the player has a total of nine on his first two cards. Again let $p_k > 1/10, 1 \leq k \leq K$, be a probability of winning that is known to occur with probability c_k. When p_k occurs, suppose the player bets a fraction f_k of his capital that The Player will receive a natural nine, and also bets a fraction f_k of his capital that The Banker will receive a natural nine. We will consider both these wagers as one bet by the player. Although the two situations are not independent, we will show that approximately the same fraction should be used on each of the two bets when both are made simultaneously, as is used when only the single bet on the banker's total is made.

If t bets are made there are t_k bets of type k with w_k^p wins and l_k^p losses for the player's hand, and w_k^b wins and l_k^b losses for the banker's hand, where $1 \leq k \leq K$. Denote the player's capital after t bets by V_t, and his initial capital by V_0.

Let the number of times in t trials such that a bet of type k is made and both The Player and The Banker have a natural nine be $w_k(b,p)$, such that The Banker has a natural nine and The Player doesn't be $w_k(b,p')$, such that The Banker does not have a natural nine and The Player does be $w_k(b',p)$, and such that neither has a natural nine be $w_k(b',p')$. Since the cards are assumed to have been shuffled randomly, by symmetry $w_k(b',p) = w_k(b,p')$.

Whether The Banker gets a natural nine is generally not independent of whether The Player does. Suppose The Banker gets a natural nine on the first deal of the shoe. The expected value of the quantity of nines in the first two cards of The Banker hand is 1/2 and of non-nines 3/2. Thus the ratio rises on the average. The chance of a Player natural nine decreases given the occurrence of a Banker natural nine, and conversely. Therefore $w_k(b,p)$ tends to be less than $w_k^b\ w_k^p$ and $w_k(b,p')$ tends to be greater than $w_k^b\ l_k^p$.

In order to determine how $w_k(b',p')$ compares with $l_k^b\ l_k^p$, we argue as follows. If The Banker does not get a natural nine on, say, the first deal of the shoe, the expected value of the quantity of nines used in the first two cards of The Banker hand will be less than 2/13. Thus The Player hand was drawn from a deck that was slightly nine-rich and so the chance of a Player natural nine is slightly increased. Therefore $w_k(b',p')$ tends to be less than $l_k^b\ l_k^p$.

Now let $p(b,p) = \lim(w_k(b,p)/t_k)$ and similarly for $p(b,p') = p(b',p)$ and $p(b',p')$. Let $p = \lim(w_k^b/t_k) = \lim(w_k^p/t_k)$ and $q = \lim(l_k^b/t_k) = \lim(l_k^p/t_k)$. Then the previous arguments show that $p(b,p) < p^2$, $p(b',p') < q^2$, and $p(b,p') > pq$. Now the expected number of natural nines in a pair of banker-player hands is $2p(b,p) + 2p(b,p')$ and the expected number in each separately is p. Since the expectation of a sum is the sum of the expectations, we have

$$2p(b,p) + 2p(b,p') = 2p = 2p(p+q) = 2p^2 + 2pq$$

and subtracting $1 = p(b,p) + 2p(b,p') + p(b',p') = p^2 + 2pq + q^2$ from each side, we find $p(b,p) - p(b',p') = p^2 - q^2$. Thus $0 < \varepsilon = p^2 - p(b,p) = q^2 - p(b,p)$ and we have $p(b,p) = p^2 - \varepsilon$, $2p(b,p') = 2pq + 2\varepsilon$, $p(b',p') = q^2 - \varepsilon$, $\varepsilon > 0$.

It is easy to see that the true probabilities $p(b,p)$, etc., do not correspond to an assumption of independence using some other probabilities p' and $q' = 1 - p'$ in place of p and q. For if this were the case, let $p' = p + \delta$, $q' = q - \delta$. Then $(p + \delta)^2 = p(b,p) = p^2 - \varepsilon$ so $2p\delta + \delta^2 = - \varepsilon < 0$, which implies $\delta < 0$. On the other hand

$(q - \delta)^2 = p(b',p') = q^2 - \varepsilon$ so $- 2q\delta + \delta^2 = - \varepsilon < 0$, which implies $\delta > 0$, a contradiction.

This shows us the effect of the assumption of independence on the $w(b,p)$, etc., and on the probabilities $p(b,p)$, etc. For simplicity, we assume independence in the next derivation.

Then we have:

$$V_t = V_0 \prod_{k=1}^{K} [(1+18f_k)^{(w_k^p w_k^b)/t_k} (1+8f_k)^{(w_k^p l_k^b)/t_k} (1+8f_k)^{(l_k^p w_k^b)/t_k} (1-2f_k)^{(l_k^p l_k^b)/t_k}]$$

$$G = \lim_{t\to\infty}[(\log_2[V_t/V_0])/t]$$

$$= \lim_{t\to\infty}(1/t) \sum_{k=1}^{K} (1/t_k)[w_k^p w_k^b \log_2(1+18f_k) + w_k^p l_k^b \log_2(1+8f_k)$$

$$+ l_k^p w_k^b \log_2(1+8f_k) + l_k^p l_k^b \log_2(1-2f_k)]$$

$$= \lim_{t\to\infty} \sum_{k=1}^{K} (t_k/t)(1/t_k^2)[w_k^p w_k^b \log_2(1+18f_k) + w_k^p l_k^b \log_2(1+8f_k)$$

$$+ l_k^p w_k^b \log_2(1+8f_k) + l_k^p l_k^b \log_2(1-2f_k)]$$

$$= \sum_{k=1}^{K} c_k [p^2 \log_2(1+18f_k) + 2pq \log_2(1+8f_k) + q^2 \log_2(1-2f_k)].$$

Therefore $\dfrac{\partial G}{\partial f_k} = c_k[18p^2/(1+18f_k) + 16pq/(1+8f_k) -2q^2/(1-2f_k)]$.

We want to determine the values of f_k which maximize G. Setting $\dfrac{\partial G}{\partial f_k} = 0$ and $q = 1-p$, we have:

$$18p^2/(1+18f_k) + 16p(1-p)/(1+8f_k) -2(1-p)^2/(1-2f_k) = 0.$$

Hence $144f_k^2 + 2(50p^2 - 90p + 13)f_k - (10p - 1) = 0$

Observe that the left side evaluated at $f_k = 0$ is negative when $p > 1/10$ therefore the equation has precisely one positive root and one negative root. By applying the quadratic formula and selecting the positive root we obtain:

$$f_k = (1/144)(-[50p^2 - 90p + 13] + 5[100p^4 - 360p^3 + 376p^2 - 36p + 1]^{1/2}).$$

We denote the optimal fraction to be bet on each of the events "the banker gets a natural nine" and "the player gets a natural nine" by f^{*BP}, so the total bet on both natural nines is $2f^{*BP}$. We denote the bet on the banker's total only, by f^{*B}. Recall that

$$p_k = (9f^{*B} + 1)/10$$

so f^{*BP} can be written as a function of f^{*B}. Table 10 gives values determined from this relationship.

TABLE 10. Comparison of f^{*B} and f^{*BP}

f^{*B}	f^{*BP}	$\frac{f^{*BP} - f^{*B}}{f^{*B}}$
.0000	.0000	-0.0%
.0100	.0100	-0.0%
.0200	.0199	-0.5%
.0300	.0298	-0.7%
.0400	.0397	-0.8%
.0500	.0494	-1.2%
.0600	.0590	-1.7%
.0700	.0686	-2.0%
.0800	.0780	-2.5%
.0900	.0873	-3.0%
.1000	.0966	-3.4%
.1100	.1057	-3.9%
.1200	.1148	-4.3%
.1300	.1237	-4.8%
.1400	.1325	-5.4%
.1500	.1412	-5.9%

One may wonder whether the functions f^{*B}, f^{*BP}, and $(f^{*BP} - f^{*B})/f^{*B}$ really are as smooth as Table 10 suggests. Conceivably, the points we chose just happened to fit smooth curves but in reality the behavior between the table values is "wild".

That these functions are smooth is a consequence of the following theorem. The theorem also has independent interest.

Theorem 1. Let $(p_1, \ldots, p_n)$ and $(p_1', \ldots, p_n')$ be probability distributions with $p_i \leq p_i'$ for $1 \leq i \leq k \leq n$ and $p_i \geq p_i'$ for $k < i \leq n$. Suppose that in one betting situation, a one-unit bet wins x_i units with probability p_i and that in a second situation, a one-unit bet wins x_i' units with probability p_i'. Suppose $x_1 \geq \ldots \geq x_k > 0 \geq x_{k+1} \geq \ldots \geq x_n \geq -1$ and $x_i' \geq x_i$, $1 \leq i \leq n$. Suppose also that if $x_i \leq 0$ or $x_i' \leq 0$ then $x_i = 0$ or -1, or $x_i' = 0$ or -1. Suppose also that for some $i_0 \leq k$, $p_{i_0}' x_{i_0}' > p_{i_0} x_{i_0}$.

(i) Then for each fixed fraction f such that $0 \leq f < 1$, $G'(f) > G(f)$, where $G(f)$ and $G'(f)$ are the associated exponential rates of growth. In particular, $G'_{max} > G_{max}$.

(ii) If some $p_i x_i < 0$, then, $f'_{max} > f_{max}$, where f_{max} and f'_{max} are the unique fractions such that $G(f_{max}) = G_{max}$ and $G'(f'_{max}) = G'_{max}$. If the mean $\mu = \Sigma p_i x_i > 0$, then there is exactly one fraction $f_c > f_{MAX}$ such that $0 < f_c < 1$ and $G(f_c) = 0$.

Proof. To prove (i), note that $G'(f) - G(f) =$

$$\sum_{i=1}^{n} p_i' \log_2(1 + x_i' f) - \sum_{i=1}^{n} p_i \log_2 (1 + x_i f) =$$

$$\sum_{i=1}^{k} \{p_i' \log_2(1 + x_i' f) - p_i \log_2(1+x_i f)\} +$$

$$\sum_{i=k+1}^{n} \{p_i \log_2(1 + x_i f)^{-1} - p_i' \log_2(1 + x_i' f)^{-1}\},$$

which is non-negative because each term of each sum is non-negative, hence $G'(f) \geq G(f)$. The condition on i_0 insures that the i_0 term is strictly positive hence $G'(f) > G(f)$.

The condition $x_n f > -1$ is needed to guarantee that the logarithms are defined.

To establish (ii), note that $dG'(f)/df - dG(f)/df =$

$$\sum_{i=1}^{n} \frac{p_i' x_i'}{1+x_i' f} - \sum_{i=1}^{n} \frac{p_i x_i}{1+x_i f} = \sum_{i=1}^{k} \left\{ \frac{p_i' x_i'}{1+x_i' f} - \frac{p_i x_i}{1+x_i f} \right\} +$$

$$\sum_{i=k+1}^{n} \left\{ \frac{p_i' x_i'}{1+x_i' f} - \frac{p_i x_i}{1+x_i f} \right\}$$

and an examination by cases shows that each term of each sum is non-negative and is strictly positive if and only if $p_i' x_i' > p_i x_i$. Thus $dG'(f)/df > dG(f)/df$.

Now $d^2G(f)/df^2 = - \sum_{i=1}^{n} \frac{p_i x_i^2}{(1+x_i f)^2} < 0$ always. Thus the equation

$dG(f)/df = 0$ has at most one solution on $[0,1)$ so there is at most one value of f which maximizes G on $[0,1)$.

If all $p_i x_i \geq 0$, $f=1$ is clearly the unique maximizing fraction for G on $[0,1]$. Suppose instead some $p_i x_i \leq 0$. Then $G(1) = -\infty$.

Since $G(0) = 0$, the maximum must occur on $[0,1)$. Therefore, since $dG'(f)/df > dG(f)/df$, $f'_{max} > f_{max}$.

Now $dG(f)/df|_{f=0} = \sum_{i=i}^{n} p_i x_i = \mu$ so $f_{max} = 0$ if $\mu \leq 0$ and $0 < f_{max} < 1$ if $\mu > 0$.

Thus the graph of G may be roughly thought of as a parabola opening downward and passing through the origin. The "vertex" is to the left of, at, or to the right of the origin, according to whether $\mu < 0$, $\mu = 0$ or $\mu > 0$.

Our assertion about table 10 follows at once from part (ii) of the theorem for it implies that f^{*B} and f^{*BP} increase monotonically as the deck becomes more nine-rich for this implies that the probability distributions for the two processes become more favorable. For $\mu > 0$, the "critical fraction" f_c is the other root of this "parabola".

We will next show that the true values of f^{*BP} are greater than those obtained from the assumption of independence, i.e. that they are greater than the values listed in Table 10, at least over the region of practical play. Thus the values in Table 10 for f^{*BP} are a lower bound for the true values, in practice. Since the values listed in Table 10 for f^{*BP} are close enough to the values of f^{*B} so that the difference can be neglected, this justifies our assumption made earlier that when the casino allows bets on both banker natural nine and player natural nine, the joint bet may be taken for all practical purposes to be twice the bet on banker natural nine only.

If we now carry through the preceding derivation without the assumption of independence, we obtain the following equation for f_k.

$$144f_k^2 + 2f_k(50p^2 - 90p + 13 - 50\varepsilon) - (10p - 1) = 0$$

This differs from the previous expression only by the additional -50ε in the coefficient of f_k. We examine the effect of this perturbation on the positive root. The unperturbed equation has the form $ax^2 + 2bx + c = 0$, and the roots are given by $x = \dfrac{-b \pm \sqrt{b^2 - ac}}{a}$.

Since only the numerator is affected by the perturbation, and $-ac > 0$ in our case, we need only consider the effect of perturbing b in the expression $\sqrt{b^2 + \Delta} - b$, where $\Delta > 0$. Now $0 < \varepsilon < p^2$ so $b = 50p^2 - 90p + 13 - 50\varepsilon > 50p^2 - 90p + 13 - 50p^2 = 13 - 90p$. Therefore $b > 0$ if $13 - 90p \geq 0$, i.e. if $p \leq 13/90$. Now $f_k = (10p - 1)/9$ so this is equivalent to $f_k \leq 4/81 \doteq 0.049$. Table 3 shows that $r \geq 5.2$ and $n \geq 20$ will insure this.

Now if $b > 0$, $\sqrt{b^2 + \Delta} - b = \Delta/(\sqrt{b^2 + \Delta} + b)$

and this obviously increases as b decreases. Thus the positive root of the perturbed equation exceeds the positive root of the unperturbed equation as we wished to show.

An analysis could be carried through for the situation where bets are being placed on all four naturals simultaneously. We believe from the previous discussion that the assumption of independence between all four natural bets will prove to be a reasonable approximation. Therefore we omit the much longer and more complex analysis of all four naturals.

It is possible to compute the precise value of ε for each (n,t) pair and then, by use of the perturbed equation, to calculate from this value of ε the precise values of f^{*BP} and of $(f^{*BP} - f^{*B})/f^{*BP}$. A similar thing could be done in principle for the various (n,t_8,t_9) situations of interest, although the calculation would be between 20 and 200 times as long.

Thus, in principle, this question can be completely settled. However, the foregoing preliminary analysis strongly suggests that the solution is the one we have stated, so it hardly seems worth the considerable additional efforts.

A more general insight into why the true values of f^{*BP} are greater than those based on the assumption of independence is provided in the next theorem. The assertion about f^{*BP} is an immediate corollary of the theorem. Roughly, the theorem asserts that if two pay-off functions F and F' have the same mean but that F' has a "smaller spread" than F, then F' is more favorable than F in the sense that the optimal fixed fraction and growth rates for F' are greater than those for F.

Definition 2. A <u>pay-off function</u> F is a probability distribution $(p_1,\ldots p_n)$ on a finite sample space and a set of numbers $(x_1,\ldots x_n)$. The expected value of the pay-off is $\Sigma p_i x_i$. Suppose $0 < a_i \leqq p_i$, $0 < a_j \leqq p_j$, $i \neq j$.

The process of forming the new pay-off function $F' = (p_1',\ldots p_n',p_{n+1}')$, $(x_1',\ldots,x_n',x_{n+1}')$ where $p_k' = p_k$ and $x_k' = x_k$, if $k \neq i,j,n+1$; $p_i' = p_i - a_i$, $x_i' = x_i$; $p_j' = p_j - a_j$, $x_j' = x_j$; and $p_{n+1}' = a_i + a_j$, $x_{n+1}' = (a_i x_i + a_j x_j)/(a_i + a_j)$, is called a <u>step</u>.

The pay-off function $F' = (p'_1, \ldots, p'_m), (x'_1, \ldots, x'_m)$ is a <u>convex contraction</u> (or simply a <u>contraction</u>) of the pay-off function $F = (p_1, \ldots, p_n), (x_1, \ldots, x_n)$ if F' can be obtained from F in a finite number of steps.

The <u>game associated with F</u> is the repeated independent trials process in which a player bets a fixed fraction f of his capital at each turn, and wins with probability p_i, x_i units per unit bet.

Observe that the mean of a contraction F' is the same as that of the original pay-off function F.

Theorem 3. If F' is a convex contraction of F, and f'^* and f^* are the fixed fractions which respectively maximize the exponential rates of growth G' and G, then $dG'/df > dG/df$. Consequently $f'^* > f^*$, $G'(f) > G(f)$ for $0 < f < 1$, and $G'_{max} > G_{max}$.

Proof. It suffices by induction to establish the theorem for a one-step contraction. From the information established in Theorem 1, it suffices to prove that $dG'/df > dG/df$. We have $dG'/df - dG/df =$

$$\frac{a_i x_i + a_j x_j}{1 + [(a_i x_i + a_j x_j)/(a_i + a_j)]f} - \left\{ \frac{a_i x_i}{1 + x_i f} + \frac{a_j x_j}{1 + x_j f} \right\} =$$

$$\left(\frac{a_i + a_j}{f}\right) \frac{\{[a_i/(a_i + a_j)]x_i + [a_j/(a_i + a_j)]x_j\} f}{1 + \{[a_i/(a_i + a_j)]x_i + [a_j/(a_i + a_j)]x_j\} f} \quad -$$

$$- \left(\frac{a_i + a_j}{f}\right)\left(\frac{[a_i/(a_i + a_j)]x_i f}{1 + x_i f} + \frac{[a_j/(a_i + a_j)]x_j f}{1 + x_j f}\right) =$$

$$[(a_i + a_j)/f] \; \{h(as + (1-a)t) - (ah(s) + (1-a)\, h(t))\} > 0,$$

where we set $s = x_i f$, $t = x_j f$, $a = a_i/(a_i + a_j)$ and h is the function $h(x) = x/(1 + x)$. The last inequality follows from the concavity of h.

The other assertions now follow from Theorem 1.

Some indication of the relation between the natural eight and natural nine bets may be helpful. They are not actually indeoendent. There is a loose "coupling". It seems plausible that if the deck is rich in nines, the probability of natural eight will be increased due to the combination $9 + 9 = 8 \pmod{10}$. Of course there are other effects of nine-richness to consider also.

Calculations show that, over the range of practical importance, that if the total number n_o of cards and the number s_o of eight are held fixed, then the probability of a natural 8 increases monotonely as the number t_o of nines is increased. However,the probability of a natural eight does not rise to the "normal" amount until n_o/t has decreased to somewhere in the neighborhood of 9.0.

Similarly, it turns out that if the deck is eight-rich, the probability of natural nine is decreased, compared with a typical distribution for the non-nine cards. The fractions might in principle be modified to account for this but the modifications appear to be too complex to be practical.

For N = 40, we will see later that the deck is sufficiently nine-rich to warrant a bet on natural nine about one seventh of the time. Similarly for eights. Simultaneous bets on both natural nine and natural eight are warranted perhaps one-twentieth of the time. Therefore the analysis of the change in betting fraction in the event all four bets on the naturals are made, analogous to the one we carried out for betting simultaneously on, say, both natural nines, is probably much less important. When bets are permitted on both natural nines they are always bet together, in equal amounts. The natural eight and natural nine bets generally occur at different times. If they do occur together, the amounts bet on each are generally rather different, so that one of them might be thought of as "dominant".

6. Average Casino advantage on all bets.

In this section we give precise values for quantities that we will use later. The complete deck probabilities for side bets that the banker has a natural are:

$$p^{8}_{416,32} = 0.09453, \text{ a house advantage of } 5.47\%.$$

$$p^{9}_{416,32} = 0.09490, \text{ a house advantage of } 5.10\%.$$

These values were computed from the equations in section 2. For a bet divided equally between the two naturals, the house advantage is therefore 5.29%.

The figures given in [7 , page 47] are erroneous. They are an advantage of 3.46% for the house in the case of a natural nine, an advantage of 6.48% for the house in the case of a natural 8, and an advantage of 4.98% for the house in case of a bet divided equally between the two naturals. Our calculations show the following complete deck probabilities.

TABLE 12

Event	Probability
The Player wins	0.446247
The Banker wins	0.458597
Tie	0.095156
	1.000000

The house advantage over The Player is 1.2351%. The house advantage over The Banker is 0.458597 x 5% - 1.2351% = 2.2930% - 1.2351% = 1.0579%, where 2.2930% is the house tax on The Banker winnings. If ties are not counted as trials, then the figures for house advantage should be multiplied by 1/0.904844, which gives a house advantage, per bet that is not a tie, over The Banker of 1.1692% and over The Player of 1.3650%. The house tax on The Banker in this situation is 2.5341%.

The figures for the house advantage given in [7, page 427] are computed as though ties are discounted but the assertion is that the figures apply to "..every hundred dealt hands in the long run ...". For reference, they are a 1.34% advantage of The Banker over The Player (compares with our 1.2351% and 1.3650% figures), a 2.53% charge on The Banker's winnings, (compares with our 2.2930% and 2.5341%), and a house advantage of 1.34% over The Player, and 1.19% over The Banker.

Thus, if the confusion over ties is clarified, the figures in [7] are quite accurate for the case where ties are not included in counting the number of trials.

7. The problem of gambler's ruin using the gambling system.

Ideally, we would like to obtain a practical algorithm for computing the probability that we will **ever** be ruined, starting with a given capital and a given set of casino rules, when betting the naturals in Nevada Baccarat - Chemin de Fer. This appears to be an extremely complicated and lengthy problem. However we will obtain upper and lower bounds for the ruin probabilities in a class of simpler situations.

To begin, let us suppose for simplicity that we have a favorable game in which the player bets a fixed fraction f of his capital V_n at each trial n and that he wins fV_n with probability p and loses fV_n with probability $q = 1 - p < p$. As we mentioned earlier, the optimal fixed fraction f, under the assumption that bets of any size are allowed, is $f = p - q$. There is in this circumstance, no probability that the player will ever be ruined.

If, instead, there is a minimum allowable bet m, then if the player's capital dwindles to $V_n < m/f$, it will no longer be possible for the player to follow the Kelly strategy. If this happens we say the player has been ruined. It is customary to think of ruin as the event that the player's capital is less than the minimum wager allowed by the player's opponent. Here we regard ruin as the event that the player's capital is so small that he cannot continue to follow his strategy. It is clear, in any case, that the existence of a minimum bet makes the probability that the player will eventually be ruined positive, no matter how great the player's initial capital. In what follows, we will continue to suppose that any size bet above the minimum is allowable.

This does not correspond to reality for us, because the casinos only allow multiples of the minimum bet, up to a certain maximum. Thus we idealize in two ways, first by allowing a continuum of bets rather than multiples of the minimum, above the minimum, and secondly by supposing that arbitrarily large bets are allowed. Each of these assumptions tends to reduce the probability of ruin below its true value. We feel that the errors which these assumptions introduce are small. Arguments to this effect will be presented later.

Ruin occurs when $V_n < m/f$ or, taking $m = 1$, when $V_n < 1/f$. Now $V_n = V_0 (1 + f)^{W_n} (1 - f)^{L_n}$ where W_n is the number of wins in n trials, L_n is the number of losses in n trials, and $W_n + L_n = n$. Thus ruin occurs when $V_0(1 + f)^{W_n} (1 - f)^{L_n} < 1/f$ or

(*) $\quad W_n \log_k(1 + f) + L_n \log_k(1 - f) < - \log_k V_0 f,$

where k is any convenient base of logarithms. This is equivalent to when absorption occurs, at the barrier $c_3 = - \log_k V_0 f$, in a one-dimensional random walk in which the particle starts at the origin and at each trial either moves left $c_2 = - \log_k (1 - f)$ units with probability q or moves right $c_1 = \log_k(1 + f)$ units with probability p. This problem has a well-known solution in the event that c_1, c_2 and c_3 are integers [Feller, Ch.XIV], and therefore in the event that c_1, c_2 and c_3 are rationals.

In general c_1, c_2 and c_3 will not be rational. However, the problem reduces to the well-known case if we merely require that $r = c_2/c_1$ be rational, for the equation $c_1 W_n - c_2 L_n < c_3$ is equivalent to $W_n - rL_n < c_3/c_1$ and if $r = a/b$, where a and b are integers and we may assume a/b is in lowest terms, we have $bW_n - aL_n < b(c_3/c_1)$. But since our particle takes integral steps, this is equivalent to absorption at c, where c is the greatest integer less than $b(c_3/c_1)$.

The condition $c_2/c_1 = a/b$ is equivalent to $-\log_k(1 - f)/\log_k(1 + f) = a/b$ or $(1 + f)^a (1 - f)^b = 1$ where $a > b > 0$. It is easy to show that there is a unique $f > 0$ which satisfies this equation, and that this f increases from 0 to 1 as a/b increases from 1 to ∞. For example, when $a/b = 0.2/0.1 = 2$, we must find the positive root of $(1 + f)^2 (1 - f) = 1$. It is $f = (-1 + \sqrt{5})/2 \doteq 0.618$.

The case where $c_2/c_1 = a/b$ is useful for the construction of examples. However, the numbers c_1 and c_2 arise from f, which is given in advance, and they are usually such that c_1/c_2 is not rational. To solve this general case, we approximate by solutions of cases where c_2/c_1 is rational. If the particle takes steps of size 1 to the right with probability p and steps of size $r_i = a_i/b_i > c_2/c_1$ to the left with probability q, then clearly the probability U_i of absorption below c_3/c_1 is larger. Similarly, if $1 < s_i = e_i/f_i < c_2/c_1$, where e_i and f_i are positive integers, and the steps to the left are instead of this size, the probability L_i of absorption is smaller. Thus, if A is the probability of absorption in the original problem, $L_i \leqq A \leqq U_i$. Further, as i increases, $U_i \downarrow U \geqq A$ and $L_i \uparrow L \leqq A$.

Note that $U = A = L$. For if $1 < a_i/b_i < c_2/c_1 < a_i/b_i + \varepsilon_i$, where ε_i is also rational, then $U_i - L_i$ is the probability that $W_n - (a_i/b_i)L_n < c_3/c_1 + \varepsilon_i$ occurs but that $W_n - (a_i/b_i)L_n < c_3/c_1$ does not occur. It is easily seen now that as $\varepsilon_i \to 0$, this probability tends to zero, hence that $U_i \to L_i$.

We illustrate these ideas for the simple game described at the beginning of this section. Suppose $p - q = f = 0.1$, $V_0 = 20$, $m = 1$. Then ruin occurs if $V_n < 1/f = 10$, or $W_n \log_{10}1.1 + L_n \log_{10}0.9 < -\log_{10}2$, which is $W_n - (1.1054^+)L_n < -(7.2725^-)$. To determine a value for U_i, we let $a_i/b_i = 10/9 = 1.1111^+$, and we have $9W_n - 10L_n < -(65.4524^-)$. This is equivalent to absorption at -66 in the associated random walk.

This problem may be solved by the methods of [Fe, XIV. 8]. However this requires first finding all the 19 roots of the characteristic equation, which in our example is $0.55s^9 + 0.45s^{-10} = 1$. The solution to the ruin problem has the form $u_z = \Sigma A_k s_k^z$ where the s_k are the roots and the A_k are arbitrary constants to be determined. Next we must use these roots and the boundary conditions [Fe, page 332] to form an array of 19 linear equations to be solved for the A_k. As the rational approximations a_i/b_i improve, the number $a_i + b_i$, which is the degree of the characteristic equation, will in general tend to infinity. Thus there is a practical limit to how closely we can estimate A.

It is easier, however, to get bounds on our U_i by the technique described in [Fe, page 334]. The characteristic equation has a unique positive root s different from 1. In our examples, $0 < s < 1$ because the mean is positive. It follows from [Fe, XIV. 8.12] that $s^{z+a_i-1} \leq U_i \leq s^z$, where z is the distance from the origin to the absorption point, which in this instance is 66. Thus $s^{75} \leq U_i \leq s^{66}$ and it remains to find s. Calculation shows that $s = 0.99000 \pm 0.00003$. Consequently $s^{75} \doteq 0.4706 \pm 0.0022$ and $s^{66} = 0.5151 \pm 0.0020$. Therefore $A \leq U_i \leq 0.5171$. Note that only the upper bound s^{66} is used here, since a lower bound for U_i doesn't imply anything about A.

To determine a value for L_i, we let $a_i/b_i = 21/19 = 1.1053^-$. We have $19W_n - 21L_n < -138.18$ so $z = 139$. The characteristic equation is $0.55s^{19} + 0.45s^{-21} = 1$. The desired root $s = 0.99499 \pm 0.00001$ whence $s^{160} \doteq 0.4484 \pm 0.0016$ and $s^{139} \doteq 0.4982 \pm 0.0014$. Therefore $A \geq L_i \geq 0.4468$, so we have found that $0.4468 \leq A \leq 0.5171$.

In the general situation where we have a pay-off function $F = (p_1,\ldots,p_n), (x_1,\ldots,x_n)$, equation (*) becomes

$$(**) \qquad \sum_{i=1}^{n} w_i \log_k(1 + x_i f) < -\log_k V_0 f, \text{ or } \sum_{i=1}^{n} c_i w_i < c.$$

As before, we approximate the probability of the (**), i.e. the probability A of ruin in the associated random walk, by determining bounds, for U_i and L_i, and hence for A. To determine a bound for L_i, we approximate by a random walk where the c_i are replaced by rational numbers r_i such that $r_i > c_i$. To determine a bound for U_i, we replace the c_i by r_i such that $r_i < c_i$. We then proceed as before.

The following tables show how the probability of ruin varies with the optimal fixed fraction f. Notice that the entries in Table 12 are virtually identical to the corresponding entries in Table 11. This may be explained as follows. The process in Table 12 of betting simultaneously on both Banker natural nine and Player natural nine is equivalent to making two successive bets on Banker natural nine. However, in contrast to the process in Table 11, instead of adjusting the value of the current capital to account for the outcome of the first bet, the value of the current capital is not adjusted until after both bets are completed.

Thus one expects the true ruin probabilities for the various situations in Table 12 to be slightly larger than the ones for the corresponding situations in Table 11. Since we don't know the true ruin probabilities, we can't compare them directly. However, since they're quite close, it's not suprising that the estimates for the lower and upper bounds are also close.

TABLE 11. The probability of ruin when betting Banker natural nine only*.

probability of winning	approximate ratio when n = 65	optimal fixed fraction	bounds on the probability of ruin A (where $1 \leq A \leq u$) where V_0 equals in minimum units											
			200		400		600		800		1000		2000	
p	r	f	l	u	l	u	l	u	l	u	l	u	l	u
.1009	11.2	.001									0.36	1.00	0.18	0.50
.1018	11.0	.002					0.30	0.83	0.22	0.62	0.18	0.50	0.09	0.25
.1036	10.5	.004			0.22	0.62	0.15	0.42	0.11	0.31	0.09	0.25	0.04	0.12
.1054	10.0	.006	0.30	0.83	0.15	0.42	0.10	0.28	0.07	0.21	0.06	0.17	0.03	0.08
.1099	9.0	.011	0.16	0.45	0.08	0.23	0.05	0.15	0.04	0.11	0.03	0.09	0.01	0.05
.1162	8.0	.018	0.10	0.28	0.05	0.14	0.03	0.09	0.02	0.07	0.02	0.06	0.01	0.03
.1234	7.0	.026	0.06	0.19	0.03	0.10	0.02	0.06	0.01	0.05	0.01	0.04	0.00	0.02
.1324	6.0	.036	0.04	0.14	0.02	0.07	0.01	0.05	0.01	0.03	0.00	0.03	0.00	0.01
.1432	5.0	.048	0.03	0.10	0.01	0.05	0.01	0.04	0.00	0.03	0.00	0.02	0.00	0.01
.1576	4.0	.064	0.02	0.08	0.01	0.04	0.00	0.03	0.00	0.02	0.00	0.02	0.00	0.01
.1747	3.0	.083	0.02	0.06	0.01	0.03	0.00	0.02	0.00	0.02	0.00	0.01	0.00	0.01
.1837	2.1	.093	0.01	0.05	0.00	0.03	0.00	0.02	0.00	0.01	0.00	0.01	0.00	0.01

*We make the approximating assumption that the continuum of bets greater than or equal to 1 unit is allowed and that there is no maximum bet.

TABLE 12. The probability of ruin when betting both Banker natural nine and Players natural nine*.

probability of winning	approximate ratio when n = 65	optimal fixed fraction	bounds on the probability of ruin A (where $1 \leq A \leq u$) where V_0 equals in minimum units: 200		400		600		800		1000		2000	
p	r	f	l	u	l	u	l	u	l	u	l	u	l	u
.1009	11.2	.001											0.18	0.50
.1018	11.0	.002									0.18	0.50	0.09	0.25
.1036	10.5	.004					0.15	0.42	0.11	0.31	0.09	0.25	0.04	0.13
.1054	10.0	.006			0.15	0.42	0.10	0.28	0.07	0.21	0.06	0.17	0.03	0.08
.1099	9.0	.011	0.16	0.46	0.08	0.23	0.05	0.15	0.04	0.11	0.03	0.09	0.01	0.05
.1162	8.0	.018	0.09	0.28	0.04	0.14	0.03	0.09	0.02	0.07	0.02	0.06	0.01	0.03
.1234	7.0	.026	0.06	0.19	0.03	0.10	0.02	0.07	0.01	0.05	0.01	0.04	0.00	0.02
.1324	6.0	.036	0.04	0.14	0.02	0.07	0.01	0.05	0.01	0.04	0.01	0.03	0.00	0.01
.1432	5.0	.048	0.03	0.11	0.01	0.06	0.01	0.04	0.00	0.03	0.00	0.03	0.00	0.01
.1576	4.0	.064	0.02	0.08	0.01	0.04	0.00	0.03	0.00	0.02	0.00	0.02	0.00	0.01
.1747	3.0	.083	0.02	0.07	0.01	0.04	0.00	0.02	0.00	0.02	0.00	0.01	0.00	0.01
.1837	2.1	.093	0.02	0.06	0.01	0.03	0.00	0.02	0.00	0.02	0.00	0.01	0.00	0.01

*We make the approximating assumption that the continuum of bets greater than or equal to 1 unit is allowed and that there is no maximum bet.

8. The effects of approximating the optimal fixed fraction.

We earlier pointed out that in actual play only multiples of the minimum bet are allowed. Therefore it is in general not possible to bet precisely the quantity $V_t f^*$. It turns out that the loss in G due to these necessary approximations to $V_t f^*$ are not serious. In the case of bets on Banker natural nine, which we now proceed to discuss in detail, we will see that when the fraction f of the current capital actually bet satisfies $0.5f^* \leq f \leq 1.5f^*$, then G satisfies $0.75G_{max} \leq G$, over the range of interest $0 \leq f^* \leq 0.13$.

Recall that for a fixed $p > 1/10$, $G = p \log_2(1 + 9f) + (1 - p) \log_2(1 - f)$ and since $p = (9f^* + 1)/10$, we have

$$G(f) = [(9f^* + 1) \log_2(1 + 9f) + 9(1 - f^*) \log_2(1 - f)]/10.$$

Also, the critical fraction $f_c > 0$ such that $G(f_c) = 0$ is the positive root of the equation $(1 + 9f)^p (1 - f)^q = 1$.

We have studied these quantities with a high-speed computer and the results follow.

TABLE 13.* The effect on G of deviating from f^*.

f^*	f_c	$f/f^*=0.1$	$f/f^*=0.3$	$f/f^*=0.5$	$f/f^*=1.0$	$f/f^*=1.5$	$f/f^*=1.7$	$f/f^*=1.9$
.01	.021	0.01	0.03	0.05	0.06	0.05	0.03	0.02
.02	.042	0.05	0.13	0.19	0.25	0.19	0.14	0.08
.03	.064	0.11	0.29	0.42	0.54	0.43	0.33	0.20
.04	.087	0.19	0.51	0.73	0.94	0.76	0.59	0.37
.05	.111	0.30	0.78	1.12	1.45	1.17	0.93	0.61
.06	.134	0.43	1.12	1.59	2.04	1.67	1.34	0.91
.07	.159	0.59	1.51	2.14	2.73	2.25	1.82	1.28
.08	.183	0.77	1.95	2.75	3.50	2.90	2.37	1.70
.09	.208	0.97	2.44	3.44	4.35	3.63	2.99	2.17
.10	.233	1.19	2.99	4.19	5.28	4.43	3.67	2.70
.11	.258	1.43	3.59	5.01	6.29	5.30	4.41	3.28
.12	.283	1.70	4.23	5.89	7.37	6.23	5.21	3.92
.13	.308	1.99	4.92	6.83	8.52	7.22	6.07	4.59
.14	.333	2.29	5.66	7.83	9.74	8.28	6.98	5.31
.15	.357	2.63	6.45	8.88	11.03	9.40	7.94	6.07

*The values in the table are of $G(f) \times 10^2$. Note that although f_c/f^* is increasing over the range studied, $0 < f^* < f_c < 1$ so eventually, as f^* increases towards 1, f_c/f^* will decrease towards 1.

Figure 2. Illustrating the effect on G of deviating from f^*.

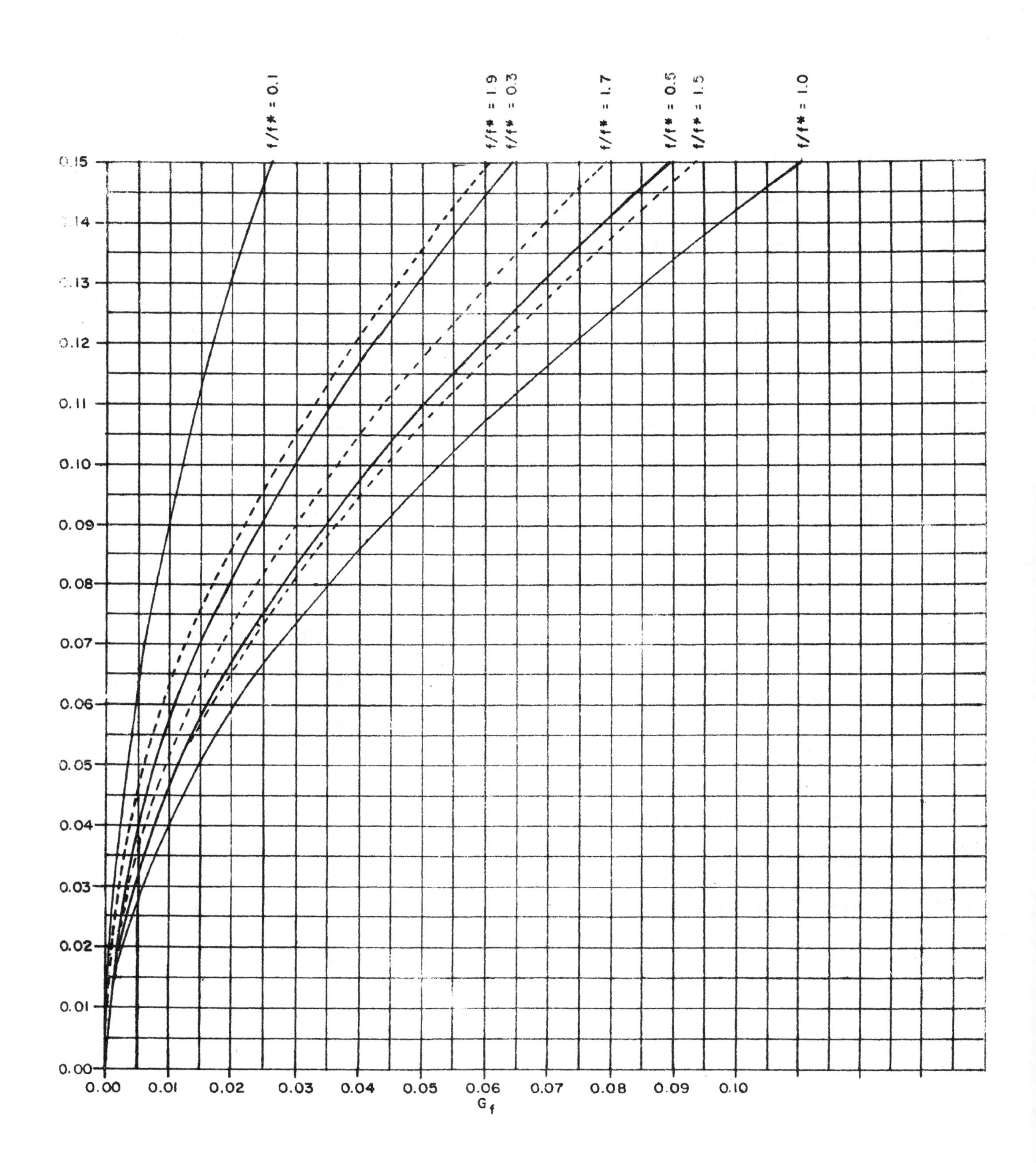

Figure 3. The relation of f_c to f^* when $0 \leq f^* \leq 0.14$

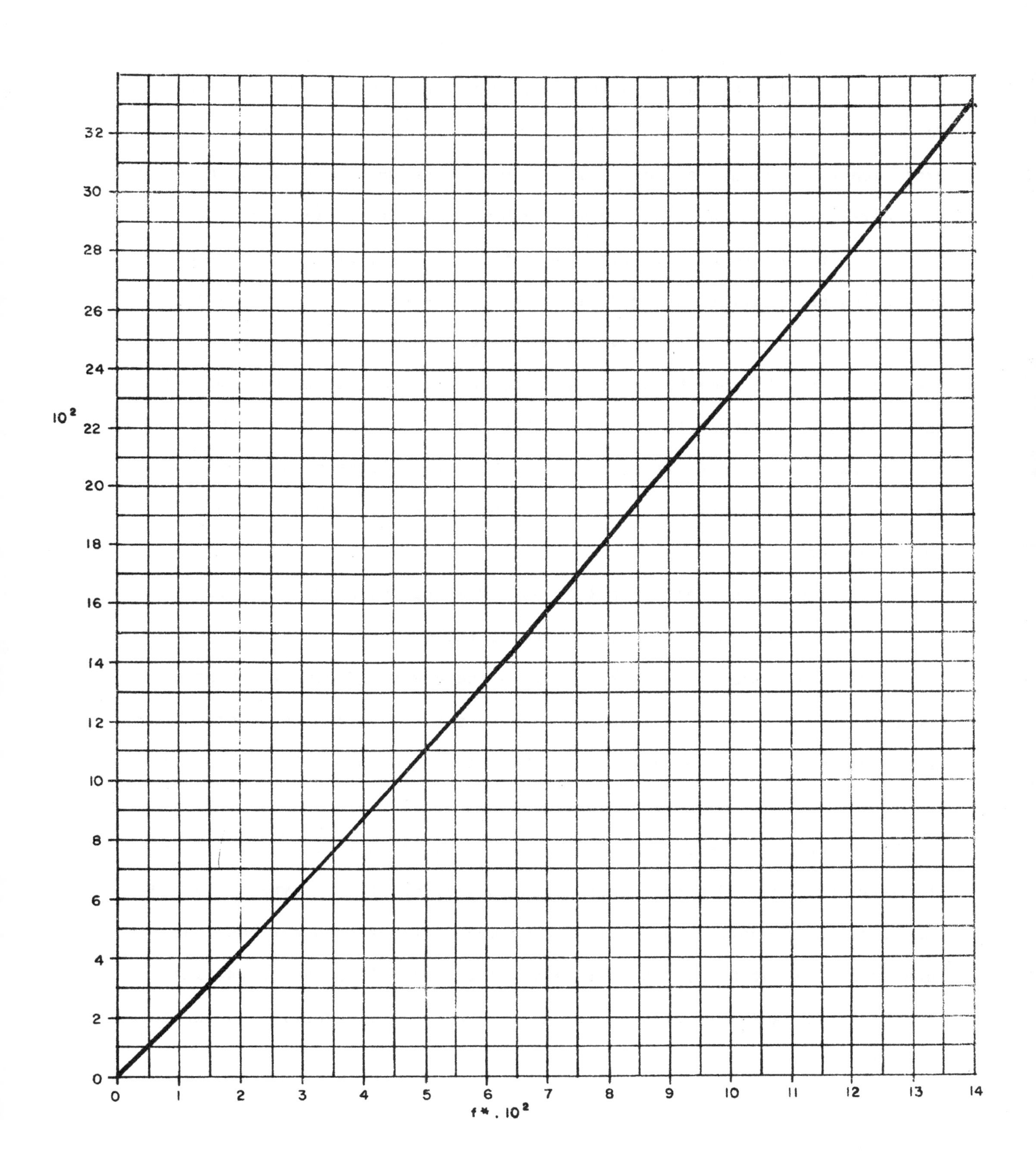

9. Practical casino play.

The difference between applied mathematics and pure mathematics seems to us to be that applied mathematics models some aspect of reality. We mean reality in the broad sense as any source of observations or experiments which may be used as a test of the mathematical model or theory (the physical world, human society, psychological data, games, etc.).

It therefore seems reasonable to test the adequacy of a piece of applied mathematics, at least in part, by how well it fits the aspect of reality it models. Often, as in our case, practical approximations to the theory must be made before it is used in practice, hence to complete the discussion we ought to know what these approximations are and how well the approximated theory "fits" reality. Therefore it seems appropriate to discuss the approximations to our theory which we have found convenient for actual casino play.

With an initial capital of $4,000 at casino A, we set aside $1,000 to cover expenses and also to cover the expected losses on the main bets. As we will see later, variations in the eight and nine richness of the deck have a neglible effect on the expectations in the main bet. Thus we can assume that those Banker bets which are made when neither natural bet is favorable loses at the usual mean rate of 1.06%. and that bets on The Player made at these times lose at the usual mean rate of 1.24%. On the average, about 80 bets in each hundred will be of this type. Assuming that they're randomized and minimum, the mean rate of loss per hour due to such bets is about 0.9m. Thus at casino A the mean loss due to waiting bets could be made less than $5 per hour. However, fluctuations about this mean also occur.

If t is the number of trials, a fluctuation of two standard deviations from the mean has an approximate magnitude of $2m\sqrt{t}$. Thus if one were to play for 60 hours, there would be about 4800 trials on the main bet, the mean loss would be less than \$300, and the fluctuation below this mean would be less than \$700 with a probability of about 98%. Thus \$1000 would probably be adequate to cover the expected losses and fluctuations in 60 hours. By installing two players betting against each other on the main bets, the fluctuations can be eliminated at the expense of doubling the mean rate of loss. Either approach, or some mixture, may be employed.

It is virtually impossible to keep close track of V_t as play progresses. We therefore decided to approximate in steps of \$500. We did this by assuming that $V_t = \$2500$ when $\$2500 \leq V_t < \$3000$, $V_t = \$3000$ when $\$3000 \leq V_t < \$3500$, etc.

We computed simplified betting tables for various values of V_t. Our capital generally stayed near enough to V_t during any one shoe that we could learn a suitable table, play, and then between shoes learn a new table if appropriate. The tables were computed from our approximate value of V_t and Tables 2-6. The results, in dollar amounts were then rounded to multiples of the minimum. Occasionally the rounding was distorted to give the tables a mnemonic pattern.

TABLE 14.* Precise amount to nearest dollar of joint bet that both the banker and the player have a natural nine (or eight).

current capital V_t →

↓ ratio	$1000	$2000	$3000	$4000	$5000
11.15	1	2	3	4	5
11.0	2	4	(5) 6	8	10
10.0	11	22	(35) 33	44	55
9.0	22	44	(65) 66	88	110
8.0	36	72	(105) 108	144	180
7.0	52	104	(155) 156	208	260
6.0	72	144	(215) 216	288	360
5.0	96	192	(290) 288	384	480
4.0	128	256	(385) 384	512	640
3.0	166	332	(500) 498	664	830
2.0	184	368	(550) 552	736	920

*The final approximate values chosen for $3000 are shown in parenthesis. Values above the maximum of $200 are shown because several players can combine bets to exceed the maximum.

We felt that \$4,000, yielding in this situation $V_0 = \$3,000$, was the minimum with which the game should be played in casino A, in order to keep the probability of ruin very small.

Due to the \$20 minimum in casino B we felt that perhaps \$10,000 was a practical minimum capital if a small probability of ruin was desired.

As it happened, we approached casino B with an effective capital of \$20,000. During the brief time we were permitted to play, we bet according to Table 15.

TABLE 15. Approximate bet sizes for V_0 = \$10,000, V_0 = \$20,000 and bets on the banker naturals only.

\$10,000 ratio	bet	\$20,000 ratio	bet
11	20	11.3	0
10.5	40	11.1	20
10	60	10.9	40
9	120	10.5	80
8	180	10	120
7	260	9	240
6	360	8	360
5	480	7	520
4	640	6	720
3	830	5	960
2	920		

10. When to stop playing in a given shoe.

We remarked earlier that favorable side-bets occur perhaps 20% of the time. In our experience shoes range from ones where more than half of all bets are favorable down to shoes where no favorable bet occur. It has proven useful in practice to have some sort of criterion by which the player knows when a shoe has become so unfavorable that the expected return on the side-bet is cancelled by the expected losses on the main bet and by the energy expended in play.

The exponential rate of growth G was originally computed using 416 cards, N of which were not used in play, where 32 of these cards were nines. Suppose this is generalized so that the exponential rate of growth $G_{n,t}$ is computed given n cards, t of which are nines and N of which are not used in play. For a given value of n we can think of G as being a "normal" rate of growth. Then define $E = G_{n,t}/G$. The quantity E can be thought of as an efficiency ratio which one can use to determine whether to play the remainder of the shoe.

We decided in practice to check E at $n = 300$, $n = 200$ and $n = 100$. In casino A, where N was on the average a little less than 40, we did this by checking to see if n/t exceeded 20. If so, we took a break until the next shoe began. This corresponds roughly to $E = 0.1$, which is the point at which we decided it was no longer worth playing.

Table 16 gives values of the ratio n/t for various choices of E, n and N.

TABLE 16. Values of the ratio n/t for various choices of E, n and N.

N = 20

E	104	143	182	221	260	299	338	377	416
.1	25.5	25.5	25.2	25.0	24.2	23.6	23.0	22.8	22.2
.2	22.0	21.7	21.2	20.8	20.4	19.8	19.1	19.0	18.6
.3	20.2	19.7	19.3	18.8	18.4	17.9	17.5	17.3	16.9
.4	19.0	18.5	18.0	17.5	17.2	16.7	16.4	16.2	15.8
.5	18.0	17.5	17.1	16.6	16.3	15.8	15.6	15.3	15.0
.6	17.1	16.8	16.4	15.9	15.6	15.2	15.0	14.7	14.5
.7	16.4	16.2	15.8	15.4	15.0	14.6	14.4	14.1	14.0
.8	15.8	15.7	15.3	14.9	14.6	14.2	14.0	13.7	13.6
.9	15.2	15.3	14.9	14.5	14.1	13.8	13.6	13.4	13.3
1.0	14.8	15.0	14.5	14.1	13.7	13.6	13.4	13.2	13.0

N = 40

E	104	143	182	221	260	299	338	377	416
.1	18.3	19.2	19.8	19.7	19.6	19.4	19.2	19.1	19.1
.2	16.8	17.5	17.6	17.7	17.5	17.5	17.2	17.1	16.9
.3	15.9	16.4	16.5	16.5	16.4	16.2	16.1	15.9	15.8
.4	15.3	15.8	15.8	15.8	15.6	15.4	15.2	15.1	15.0
.5	14.8	15.3	15.3	15.2	15.0	15.0	14.8	14.7	14.5
.6	14.6	14.9	14.8	14.7	14.6	14.4	14.3	14.2	14.1
.7	14.3	14.5	14.5	14.4	14.3	14.1	14.0	13.8	13.7
.8	14.1	14.2	14.2	14.1	14.0	13.7	13.7	13.5	13.4
.9	13.8	13.9	13.9	13.9	13.7	13.4	13.4	13.3	13.2
1.0	13.6	13.6	13.6	13.6	13.5	13.2	13.2	13.1	13.0

N = 60

E	104	143	182	221	260	299	338	377	416
.1	15.1	16.5	17.2	17.5	17.6	17.6	17.6	17.6	17.6
.2	14.4	15.4	16.0	16.1	16.2	16.2	16.2	16.2	16.0
.3	13.8	14.8	15.2	15.3	15.5	15.4	15.4	15.3	15.2
.4	13.4	14.4	14.7	14.8	14.8	14.6	14.6	14.6	14.6
.5	13.1	14.1	14.3	14.5	14.4	14.2	14.2	14.2	14.2
.6	13.0	13.8	14.0	14.1	14.2	14.0	14.0	13.9	13.9
.7	12.9	13.6	13.8	13.8	13.9	13.7	13.7	13.7	13.6
.8	12.8	13.4	13.6	13.6	13.6	13.4	13.4	13.4	13.4
.9	12.7	13.2	13.4	13.4	13.4	13.2	13.2	13.2	13.2
1.0	12.6	13.0	13.2	13.2	13.2	13.0	13.0	13.0	13.0

11. The distribution of f^* for various N.

It is of interest to have some idea of the probability distribution of f^*. For example, we stated earlier without proof that about 80% of all bets were waiting bets. Using this figure we estimated the rate of loss due to the waiting bets. We now justify this estimate.

Table 17 describes the distribution of f^* for various N, assuming that side bets are made only on the Banker natural nine. For example, when $N = 40$, the probability that $f^* \geq 0.001$ is 0.1463. For bets on natural eight we would expect the table corresponding to Table 17 to have very slightly larger entries in the first row and very slightly smaller entries in all the other rows. Thus, if bets on natural 8 and natural 9 each occurred with uniform probability throughout the shoe, and were independent of each other, we would expect the probability of a side bet on at least one of the naturals to be about $2x(0.146) - (0.146)^2 = 0.271$.

Actually, neither of these assumptions are true. To see that bets on the naturals are not uniformly distributed, recall that there are no favorable side bets during the first few coups of a shoe. Also the probability of a favorable side bet increases monotonely as we proceed through the shoe. The effect is to decrease the figure sharply below 0.271.

The two side bets are not independent of each other. If the shoe is rich in one type of card, it is less likely than usual to be rich in the other type of card. Conversely, if it is poor in one type of card, it is more likely to be rich in the other type. The effect of this is to increase the figure above 0.271. However we feel this effect is considerably smaller than the other one. Our attempts to estimate the magnitudes of the two effects yielded our rough estimate of 20% for the frequency of favorable side bets. Clearly

TABLE 17. The distribution of f^* for various N, when the Banker natural nine is bet.

Probability that	N = 10	N = 20	N = 40	N = 60	N = 80	N = 100
$.000 \leq f^* < .001$	.8352	.8415	.8537	.8656	.8773	.8888
$.001 \leq f^* < .002$	.0219	.0225	.0223	.0224	.0223	.0219
$.002 \leq f^* < .003$	.0183	.0178	.0187	.0187	.0183	.0178
$.003 \leq f^* < .004$	.0155	.0159	.0154	.0152	.0147	.0144
$.004 \leq f^* < .005$	.0130	.0133	.0134	.0126	.0125	.0117
$.005 \leq f^* < .006$	.0102	.0105	.0111	.0112	.0101	.0093
$.006 \leq f^* < .007$	.0106	.0099	.0091	.0082	.0080	.0072
$.007 \leq f^* < .008$	.0080	.0082	.0076	.0072	.0065	.0060
$.008 \leq f^* < .009$	.0069	.0064	.0063	.0062	.0056	.0048
$.009 \leq f^* < .010$	.0063	.0065	.0058	.0053	.0047	.0039
$.010 \leq f^* < .015$	.0205	.0197	.0184	.0159	.0130	.0103
$.015 \leq f^* < .020$	.0118	.0115	.0096	.0070	.0049	.0032
$.020 \leq f^* < .025$	.0069	.0064	.0044	.0029	.0017	.0008
$.025 \leq f^* < .030$	.0047	.0042	.0024	.0012	.0005	.0002
$.030 \leq f^* < .040$	.0052	.0036	.0017	.0006	.0002	.0001
$.040 \leq f^* < .050$	.0027	.0015	.0003	.0001	.0000	.0000
$.050 \leq f^* < .060$	.0013	.0005	.0001	.0000	.0000	.0000
$.060 \leq f^* < .070$	.0007	.0002	.0000	.0000	.0000	.0000
$.070 \leq f^* < .080$	.0003	.0001	.0000	.0000	.0000	.0000
$.080 \leq f^* < .090$	.0002	.0000	.0000	.0000	.0000	.0000
$.090 \leq f^* < 1.000$	.0000	.0000	.0000	.0000	.0000	.0000

if a more precise value is needed, we can determine it empirically by machine simulation of the play of a large number of hands.

To illustrate another use of Table 17, suppose the player's initial capital allows bets whenever $f^* \geq 0.001$. If at a later time the player's current capital is sharply reduced, so that he may bet only when $f^* \geq 0.002$, say, then his growth coefficient will be reduced. Table 17 may be used to roughly estimate this reduction in growth coefficient.

Table 17 may also be used to give a rough estimate of the entire growth coefficient. This was done originally, with an approximate version of Table 17, before the main calculations of the paper were undertaken. This estimate of the growth coefficient gave us advance assurance that our playing strategy would have an appreciable yield.

12. When the side bet on Banker natural nine is better than the main bets.

We wish to determine when, from a growth point of view, one should make the side bet on natural nine rather than the main bet.

For the side bet

$$G = p \log_2(1 + 9f) + (1 - p) \log_2(1 - f)$$

For the main bet

$$G_B = (0.45860) \log_2(1 + 0.95f) + (0.44625) \log_2(1 - f)$$

$$G_P = (0.44625) \log_2(1 + f) + (0.45860) \log_2(1 - f)$$

For this discussion f is defined as m/V_t where m is the minimum bet allowed and V_t is the current capital. When $G = G_B$ we have:

$$p \log_2(1 + 9f) + (1 - p) \log_2(1 - f) = (0.45860) \log_2(1 + 0.95f) + (0.44625) \log_2(1 - f)$$

which yields

$$p = \frac{(0.45860) \ln(1 + 0.95f) - (0.55375) \ln(1 - f)}{\ln(1 + 9f) - \ln(1 - f)}$$

where $\ln = \log_e$.

TABLE 18. When a (minimum, unfavorable) side bet on Banker natural nine gives the same rate of growth as an equal bet on The Banker or The Player.

f	Value of p when $G = G_B$	Value of p when $G = G_p$
0.0001	0.09898	0.09880
0.0005	0.09914	0.09896
0.0010	0.09934	0.09916
0.0015	0.09954	0.09936
0.0020	0.09974	0.09956
0.0030	0.10014	0.09996
0.0040	0.10054	0.10035
0.0050	0.10094	0.10075
0.0060	0.10133	0.10114

This formula yields the value of p such that $G = G_B$. Denote this value of p by p_B. Similarly, when $G = G_P$ we have

$$p = \frac{(0.44625)\ln(1 + f) - (0.54140)\ln(1 - f)}{\ln(1 + 9f) - \ln(1 - f)}$$

This formula yields the value of p such that $G = G_P$. Denote this value by p_P.

Table 18 gives selected values p_B and p_P over the range $0.0001 \leq f \leq 0.0060$. At casino A this corresponds to $\$833 \leq V_t \leq \$50,000$. At casino B it corresponds to $\$3,333 \leq V_t \leq \$200,000$.

The functions $p_B(f)$ and $p_P(f)$ are within 1.33% of the value 0.100 over this range. Thus $p_B = p_P = 0.100$ seems like a reasonable approximation in practice.

Suppose we use the criterion $p_B = p_P \geq 0.1$ to place our minimum bet on Banker natural nine rather than on the main bet. If we set $p_{n,t} = 0.1$ and solve equation (6) for r, we have, setting $1/(n - 1) = \varepsilon$,

$$509r^2 - (6380 + 7020\varepsilon)\, r + 7020(1 + \varepsilon) = 0$$

whence

$$r = \frac{(3190 + 3510\varepsilon) \pm \sqrt{(3190 + 3510\varepsilon)^2 - 1018 \times 3510(1 + \varepsilon)}}{509}$$

The solution r^+ corresponding to the positive choice of sign is of primary interest. We find that $r^+(0) = 11.316$, corresponding to $n = \infty$. Also $r^+(1/10) = 12.721$, corresponding to $n = 11$. Since $dr^+/d\varepsilon > 0$ for $\varepsilon \geq 0$, r^+ increases monotonely as n decreases.

Table 19 gives selected values of $r^+(\varepsilon)$ and $r^-(\varepsilon)$. For a given n, the Banker natural nine bet should be preferred over the main bets when $r^-(\varepsilon) \leq r \leq r^+(\varepsilon)$.

TABLE 19. When Banker natural nine is to be preferred over the main bets, using the criterion $p_B = p_P > 0.1$.

n	r^+	r^-
∞	11.32	1.22
400	11.35	1.22
350	11.36	1.22
300	11.36	1.22
250	11.37	1.22
200	11.39	1.22
150	11.41	1.22
100	11.46	1.22
75	11.51	1.21
50	11.60	1.21
40	11.68	1.21
30	11.80	1.21
20	12.06	1.20
11	12.72	1.19
10	12.88	1.19
9	13.07	1.19
8	13.32	1.18
7	13.65	1.18
6	14.12	1.17
5	14.82	1.16

<u>The main bets.</u> Our initial interest in Baccarat - Chemin de Fer arose from its vague similarity to the successfully analyzed casino game of Blackjack. We attempted to determine whether or not the abnormal composition of the shoe, which arises as successive coups are dealt, gives rise to fluctuations in the expectations of the Banker and Players bets which are sufficient to overcome the house edge. We shall see that this occasionally happens but that the fluctuations do not seem to be either large enough or frequent enough to be the basis of a practical winning strategy.

As in Blackjack, we first considered the effect on the Banker and Player bets of varying the quantity of cards of a single numerical value. Table 13 shows the results for various special decks. Decks in which the number of cards having one of the values 1 through 9 exceeds 32, or where the number of 0 valued cards exceeds 128, cannot arise in play, but they give insight into the quantities we are studying.

TABLE 20. Dependence of main bet probabilities on deck composition

Composition of deck: number of cards of indicated value; no entry means the number is normal: 0	1	2	3	4	5	6	7	8	9	The Player wins	The Banker wins	tie	The Player advantage	The Banker advantage
quarter deck										0.44517	0.46036	0.09448	-1.519%	-0.783%
half deck										0.44689	0.46041	0.09270	-1.352%	-0.9501%
one deck										0.44676	0.45962	0.09362	-1.286%	-1.0117%
two decks										0.44651	0.45907	0.09442	-1.257%	-1.0389%
four decks										0.44634	0.45876	0.09490	-1.242%	-1.052%
six decks										0.44628	0.45865	0.09507	-1.237%	-1.056%
eight decks (normal)										0.44625	0.45860	0.09516	-1.235%	-1.058%
64										0.44378	0.45678	0.09944	-1.300%	-0.984%
96										0.44500	0.45782	0.09718	-1.283%	-1.007%
127										0.44621	0.45858	0.09521	-1.237%	-1.056%
129										0.44628	0.45862	0.09510	-1.233%	-1.060%
	16									0.44574	0.45884	0.09542	-1.310%	-0.984%
	24									0.44600	0.45872	0.09528	-1.272%	-1.022%
	31									0.44622	0.45861	0.09517	-1.240%	-1.054%
	33									0.44628	0.45858	0.09514	-1.231%	-1.062%
		31								0.44623	0.45864	0.09129	-1.241%	-1.053%
		33								0.44626	0.45856	0.09518	-1.230%	-1.063%
			31							0.44622	0.45864	0.09513	-1.242%	-1.051%
			33							0.44627	0.45855	0.09518	-1.228%	-1.065%

Composition of deck: number of cards of indicated value; no entry means the number is normal														
0	1	2	3	4	5	6	7	8	9	The Player wins	The Banker wins	tie	The Player advantage	The Banker advantage
				31						0.44620	0.45867	0.09512	-1.247%	-1.046%
				33						0.44629	0.45852	0.09519	-1.223%	-1.080%
					31					0.44630	0.45857	0.09513	-1.227%	-1.076%
					33					0.44619	0.45863	0.09519	-1.244%	-1.050%
						31				0.44637	0.45861	0.09503	-1.224%	-1.069%
						33				0.44613	0.45859	0.09529	-1.246%	-1.047%
							31			0.44635	0.45862	0.09504	-1.227%	-1.066%
							33			0.44614	0.45858	0.09528	-1.243%	-1.050%
64	16	16	16	16	16	16	16	24	16	0.44626	0.45948	0.09427	-1.322%	-0.975%
								33		0.44625	0.45866	0.09509	-1.240%	-1.053%
									33	0.44626	0.45863	0.09511	-1.238%	-1.056%
64	16	16	16	16	16	16	16	16	24	0.44636	0.45911	0.09453	-1.275%	-1.020%
64	16	16	16	16	16	16	16	16	32	0.44619	0.45917	0.09464	-1.298%	-0.998%

With the exception of situations where n is quite small, it rarely happens that the relative richness of a particular card in the deck is less than half of normal (normal richness 1/13 for the cards 1 through 9 and 4/13 for zero) or greater than twice normal. Table 20 suggests strongly that over this range and even well beyond it, the casino has the advantage on both The Banker and The Player bets.

We next inquired as to whether, if one were able to analyze small n situations perfectly (e.g. the player might receive radioed instructions from a computer), there were appreciable fluctuations in advantages a significant part of the time. We selected 27 sets of 13 cards each, each set drawn randomly from eight complete decks, and computed the significant quantities. The results, arranged in order of decreasing player advantage, appear in Table 21.

The Table shows that even in an extreme end of the shoe situation where only 13 cards remain, the frequency and magnitude of favorable bets on The Banker and The Player is negligible. This seems to settle negatively the long standing question about whether or not the various forms of Baccarat can be beaten in practice by any card-counting techniques whatsoever.

TABLE 21. Main bet probabilities for randomly chosen 13 card subsets.

Composition of deck: number of cards of indicated value; no entry means the number is normal															
0	1	2	3	4	5	6	7	8	9	Total	The Player wins	The Banker wins	tie	The Player advantage	The Banker advantage
2	3	2	3	1	1	0	0	1	0	13	0.45901	0.42676	0.11423	3.226%	-5.359%
3	1	1	2	2	1	1	0	1	1	13	0.44816	0.45008	0.10176	-0.192%	-2.058%
4	0	1	4	0	2	0	0	1	1	13	0.45738	0.45952	0.08310	-0.215%	-2.083%
7	2	1	1	0	0	1	0	0	1	13	0.46376	0.46656	0.06969	-0.280%	-2.052%
7	1	3	0	0	0	0	2	0	0	13	0.45592	0.45901	0.08506	-0.309%	-1.986%
4	0	1	1	3	1	2	1	0	0	13	0.44775	0.45162	0.10063	-0.386%	-1.872%
7	0	2	0	0	0	1	1	1	1	13	0.46888	0.47338	0.05774	-0.449%	-1.918%
4	2	1	0	0	0	2	1	0	3	13	0.44779	0.45525	0.09696	-0.747%	-1.530%
2	1	1	0	3	1	2	0	1	2	13	0.44971	0.45889	0.09140	-0.917%	-1.377%
4	2	1	2	0	2	0	0	1	1	13	0.45533	0.46455	0.08012	-0.922%	-1.401%
7	1	1	1	2	0	1	0	0	0	13	0.45278	0.46311	0.08410	-1.033%	-1.283%
7	1	0	1	0	1	0	2	1	0	13	0.45445	0.46488	0.08067	-1.043%	-1.282%
2	0	1	3	1	0	3	1	1	1	13	0.44060	0.45139	0.10802	-1.079%	-1.178%
2	2	1	0	1	3	0	2	0	2	13	0.44127	0.45296	0.10577	-1.169%	-1.095%
6	0	1	2	0	0	1	2	1	0	13	0.45646	0.46887	0.07467	-1.241%	-1.103%
3	1	1	3	1	0	2	1	1	0	13	0.44157	0.45428	0.10415	-1.270%	-1.001%

Composition of deck: number of cards of indicated value; no entry means the number is normal											The Player wins	The Banker wins	tie	The Player advantage	The Banker advantage
0	1	2	3	4	5	6	7	8	9	Total					
2	0	1	1	2	0	1	2	1	3	13	0.44264	0.45547	0.10189	-1.283%	-0.994%
5	0	0	2	2	0	1	1	0	2	13	0.44424	0.45709	0.09866	-1.284%	-1.001%
2	2	1	2	1	1	1	1	0	2	13	0.44236	0.45556	0.10209	-1.320%	-0.958%
5	2	2	0	1	2	0	1	0	0	13	0.44371	0.45773	0.09856	-1.402%	-0.887%
3	1	1	0	1	1	0	2	2	2	13	0.44059	0.45549	0.10393	-1.489%	-0.789%
2	0	0	1	3	1	1	1	3	1	13	0.44629	0.46149	0.09221	-1.520%	-0.788%
5	1	1	1	1	0	2	1	1	0	13	0.44666	0.46390	0.08945	-1.724%	-0.596%
2	1	0	1	2	1	1	3	1	1	13	0.44220	0.45948	0.09832	-1.728%	-0.569%
5	0	1	0	1	2	1	1	0	2	13	0.44253	0.46369	0.09378	-2.116%	-0.202%
2	0	1	1	1	0	2	1	3	2	13	0.43892	0.46199	0.09908	-2.307%	-0.003%
2	1	1	1	0	3	1	2	1	1	13	0.43816	0.46244	0.09940	-2.427%	+0.115%

14. Applications to Blackjack.

The Blackjack results of [8] and [10] are based largely on a computer program (unpublished) which does the following. Given any collection of cards, the program computes to a "high degree of approximation" the best player strategy in all situations. It also computes the player's corresponding mathematical expectation to within about 0.3% (this figure can be improved considerably by simple program revisions). Typical rules [10, Chapter III] are assumed. The program can be altered readily to fit the variations in casino rules. In particular, it is noted that subsets of cards which are ten-rich tend to favor the player and decks which are ten-poor tend to favor the casino.

Casino Blackjack is generally played with either one, two or four decks. As in Baccarat, used cards are discarded and successive rounds come from a depleted deck (or decks). A winning system may thus be based on counting tens, which we denote by t, and non-tens which we denote by n. The computer program readily determines the player advantage for each pair of numbers (t,n).

The situation at this point is wholly analogous to that of the Baccarat side bet. We use the Kelly system for betting, for the same reasons given in section 4 for the Baccarat side bet. The computations of the optimal betting fraction and the corresponding growth rate follow the same steps. A discussion of gambler's ruin imitating that in section 7 offers no new difficulties.

In Blackjack also, it is generally possible at best to approximate the optimal betting fraction. The reasons are the same as for the Baccarat side bet and the analysis is like that in section 8.

The analysis in section 10 may be carried over to Blackjack. It is of little practical interest for the prevalent one-deck game. However, the information is quite useful in the less frequent two and four deck games. An analysis like that in section 11 for Blackjack gives information much like that obtained for Baccarat.

The above applications of the Baccarat results to Blackjack present no new difficulties. The calculations are of no greater difficulty and generally considerably simpler.

References

1. Boll, Marcel, La Chance et les Jeux de Hasard, Librairie Larousse, Paris, 1936.
2. Feller, William, An Introduction to Probability Theory and Its Applications Vol. I. Wiley, New York (1957).
3. Frey, Richard L., According to Hoyle. Fawcett Publications, Inc., Greenwich, Conn. (1957).
4. Kelly, J. L., A new interpretation of information rate. Bell System Tech. J., Vol. 35, 917-926 (1956).
5. Kemeny, J. G., and Snell, J. L., Game-Theoretic Solution of Baccarat. Amer. Math. Monthly, Vol. LXIV, No. 7, 465-469, August-September, 1957.
6. Le Myre, Georges, Le Baccara. Hermann, Paris, 1935.
7. Scarne, John, Scarne's Complete Guide to Gambling. Simon and Shuster, Inc. New York (1961).
8. Thorp, Edward O., A favorable strategy for Twenty-One. Proc. Nat. Acad. Sci. 47, 110-112 (1961).
9. ————————, A Prof Beats the Gamblers, The Atlantic Monthly, June 1962.
10. ————————, Beat the Dealer. Blaisdell, New York (1962).

CPSIA information can be obtained at www.ICGtesting.com
Printed in the USA
BVOW10s0048020814

361358BV00016B/89/P

9 781626 549456